Student Name:
Pre-Test Score:
Date Started:
Date Completed:
Final Test Score:

MULTIPLICATION

4th Grade

SAREEDO

Pre-University MATH WORKBOOKS

1. Are You Ready 1?

1)

312	245	175	469
+ 90	+ 57	+ 88	+ 57

2)

152	524	617	549
+967	+258	+686	+798

3)

712	274	176	493
+565	+256	+688	+709

4)

619	214	674
+506	+546	+673

5)

5712	2864	8796
+5659	+2567	+6654

2) Are You Ready 2?

1) $\begin{array}{r} 95 \\ -\ 36 \\ \hline \end{array}$ $\begin{array}{r} 47 \\ -\ 03 \\ \hline \end{array}$ $\begin{array}{r} 74 \\ -38 \\ \hline \end{array}$ $\begin{array}{r} 99 \\ -59 \\ \hline \end{array}$

2) $\begin{array}{r} 52 \\ -\ 39 \\ \hline \end{array}$ $\begin{array}{r} 54 \\ -51 \\ \hline \end{array}$ $\begin{array}{r} 87 \\ -69 \\ \hline \end{array}$ $\begin{array}{r} 48 \\ -19 \\ \hline \end{array}$

3) $\begin{array}{r} 520 \\ -\ 346 \\ \hline \end{array}$ $\begin{array}{r} 354 \\ -271 \\ \hline \end{array}$ $\begin{array}{r} 781 \\ -583 \\ \hline \end{array}$ $\begin{array}{r} 716 \\ -494 \\ \hline \end{array}$

4) $\begin{array}{r} 720 \\ -\ 346 \\ \hline \end{array}$ $\begin{array}{r} 300 \\ -271 \\ \hline \end{array}$ $\begin{array}{r} 766 \\ -583 \\ \hline \end{array}$

5) $\begin{array}{r} 837 \\ -654 \\ \hline \end{array}$ $\begin{array}{r} 837 \\ -674 \\ \hline \end{array}$ $\begin{array}{r} 716 \\ -464 \\ \hline \end{array}$

3) Multiplication Facts (0 and 1)

1)	0×1	0×0	0×3
2)	0×4	0×5	0×6
3)	0×7	0×8	0×9
4)	0×10	0×11	0×10
5)	1×0	2×0	3×0
6)	4×0	5×0	6×0
7)	7×0	8×0	9×0
8)	10×0	11×0	12×0

1)	1×1	1×2	1×3
2)	1×4	1×5	1×6
3)	1×7	1×8	1×9
4)	1×10	1×11	1×12
5)	1×1	2×1	3×1
6)	4×1	5×1	6×1
7)	7×1	8×1	9×1
8)	10×1	11×1	12×1

4) Multiplication Facts (2 and 3)

1)	2×1	2×2	2×3
2)	2×4	2×5	2×6
3)	2×7	2×8	2×9
4)	2×10	2×11	2×12
5)	1×2	2×2	3×2
6)	4×2	5×2	6×2
7)	7×2	8×2	9×2
8)	10×2	11×2	12×2

1)	3×1	3×2	3×3
2)	3×4	3×5	3×6
3)	3×7	3×8	3×9
4)	3×10	3×11	3×12
5)	1×3	2×3	3×3
6)	4×3	5×3	6×3
7)	7×3	8×3	9×3
8)	10×3	11×3	12×3

5) Multiplication Facts (4 and 5)

1)	4 × 1	4 × 2	4 × 3
2)	4 × 4	4 × 5	4 × 6
3)	4 × 7	4 × 8	4 × 9
4)	4 × 10	4 × 11	4 × 12
5)	1 × 4	2 × 4	3 × 4
6)	4 × 4	5 × 4	6 × 4
7)	7 × 4	8 × 4	9 × 4
8)	10 × 4	11 × 4	12 × 4

1)	5 × 1	5 × 2	5 × 3
2)	5 × 4	5 × 5	5 × 6
3)	5 × 7	5 × 8	5 × 9
4)	5 × 10	5 × 11	5 × 12
5)	1 × 5	2 × 5	3 × 5
6)	4 × 5	5 × 5	6 × 5
7)	7 × 5	8 × 5	9 × 5
8)	10 × 5	11 × 5	12 × 5

6) Multiplication Facts (6 and 7)

1)	6×1	6×2	6×3
2)	6×4	6×5	6×6
3)	6×7	6×8	6×9
4)	6×10	6×11	6×12
5)	1×6	2×6	3×6
6)	4×6	5×6	6×6
7)	7×6	8×6	9×6
8)	10×6	11×6	12×6

1)	7×1	7×2	7×3
2)	7×4	7×5	7×6
3)	7×7	7×8	7×9
4)	7×10	7×11	7×12
5)	1×7	2×7	3×7
6)	4×7	5×7	6×7
7)	7×7	8×7	9×7
8)	10×7	11×7	12×7

7) Multiplication Facts (8 and 9)

1)	8×1	8×2	8×3
2)	8×4	8×5	8×6
3)	8×7	8×8	8×9
4)	8×10	8×11	8×12
5)	1×8	2×8	3×8
6)	4×8	5×8	6×8
7)	7×8	8×8	9×8
8)	10×8	11×8	12×8

1)	9×1	9×2	9×3
2)	9×4	9×5	9×6
3)	9×7	9×8	9×9
4)	9×10	9×11	9×12
5)	1×9	2×9	3×9
6)	4×9	5×9	6×9
7)	7×9	8×9	9×9
8)	10×9	11×9	12×9

8) A Multiplication Facts (9 nd 10)

1)	9 × 1	9 × 2	9 × 3
2)	9 × 4	9 × 5	9 × 6
3)	9 × 7	9 × 8	9 × 9
4)	9 × 10	9 × 11	9 × 12
5)	1 × 9	2 × 9	3 × 9
6)	4 × 9	5 × 9	6 × 9
7)	7 × 9	8 × 9	9 × 9
8)	10 × 9	11 × 9	12 × 9

1)	10 × 1	10 × 2	10 × 3
2)	10 × 4	10 × 5	10 × 6
3)	10 × 7	10 × 8	10 × 9
4)	10 × 10	10 × 11	10 × 12
5)	1 × 10	2 × 10	3 × 10
6)	4 × 10	5 × 10	6 × 10
7)	7 × 10	8 × 10	9 × 10
8)	10 × 10	11 × 10	12 × 10

9) Multiplication Facts (11 nd 12)

1)	11 × 1	11 × 2	11 × 3
2)	11 × 4	11 × 5	11 × 6
3)	11 × 7	11 × 8	11 × 9
4)	11 × 10	11 × 11	11 × 12
5)	1 × 11	2 × 11	3 × 11
6)	4 × 11	5 × 11	6 × 11
7)	7 × 11	8 × 11	9 × 11
8)	10 × 11	11 × 11	12 × 11

1)	12 × 1	12 × 2	12 × 3
2)	12 × 4	12 × 5	12 × 6
3)	12 × 7	12 × 8	12 × 9
4)	12 × 10	12 × 11	12 × 12
5)	1 × 12	2 × 12	3 × 12
6)	4 × 12	5 × 12	6 × 12
7)	7 × 12	8 × 12	9 × 12
8)	10 × 12	11 × 12	12 × 12

10) Multiplication Facts (Mixed Practice 1)

1)	8×1	6×2	5×3
2)	7×4	4×5	8×6
3)	8×3	4×4	5×5
4)	3×3	6×6	7×7
5)	8×8	9×8	9×9
6)	4×7	8×0	2×9
7)	7×8	8×8	9×3
8)	10×8	11×8	12×8

1)	7×9	5×3	8×2
2)	6×4	4×8	8×6
3)	9×7	2×8	3×6
4)	8×7	5×11	9×12
5)	1×9	5×6	3×9
6)	4×9	0×9	4×3
7)	7×9	8×9	9×6
8)	10×12	11×9	10×9

11) Multiplication Facts (Mixed Practice 2)

1)	8×6	6×3	5×4
2)	7×5	4×9	8×3
3)	8×7	8×9	8×5
4)	9×0	7×6	8×7
5)	9×4	8×6	9×6
6)	4×7	7×0	2×9
7)	7×8	8×8	9×3
8)	10×9	1[illegible] × 1	12×8

1)	7×9	5×3	8×2
2)	6×4	4×8	8×6
3)	4×7	7×8	3×6
4)	8×7	9×11	9×12
5)	1×9	8×6	3×9
6)	4×9	0×9	5×3
7)	9×9	6×9	9×6
8)	10×12	12×12	8×9

Wow! You have finished all that in no time! Great work.

Now Time to relax and color Mr. Fox in 6 different colors!

12) Multiply 2-Digits by 0, 1, 2, 3

Name:	Data:	Score:

$$\begin{array}{r} 28 \\ \times\ 0 \\ \hline \end{array}$$ $$\begin{array}{r} 37 \\ \times\ 1 \\ \hline \end{array}$$ $$\begin{array}{r} 46 \\ \times\ 2 \\ \hline \end{array}$$ $$\begin{array}{r} 55 \\ \times\ 2 \\ \hline \end{array}$$

$$\begin{array}{r} 64 \\ \times\ 3 \\ \hline \end{array}$$ $$\begin{array}{r} 73 \\ \times\ 2 \\ \hline \end{array}$$ $$\begin{array}{r} 82 \\ \times\ 3 \\ \hline \end{array}$$ $$\begin{array}{r} 91 \\ \times\ 1 \\ \hline \end{array}$$

$$\begin{array}{r} 94 \\ \times\ 2 \\ \hline \end{array}$$ $$\begin{array}{r} 85 \\ \times\ 2 \\ \hline \end{array}$$ $$\begin{array}{r} 83 \\ \times\ 0 \\ \hline \end{array}$$ $$\begin{array}{r} 79 \\ \times\ 2 \\ \hline \end{array}$$

$$\begin{array}{r} 78 \\ \times\ 2 \\ \hline \end{array}$$ $$\begin{array}{r} 67 \\ \times\ 3 \\ \hline \end{array}$$ $$\begin{array}{r} 60 \\ \times\ 2 \\ \hline \end{array}$$ $$\begin{array}{r} 56 \\ \times\ 2 \\ \hline \end{array}$$

$$\begin{array}{r} 94 \\ \times\ 2 \\ \hline \end{array}$$ $$\begin{array}{r} 63 \\ \times\ 2 \\ \hline \end{array}$$ $$\begin{array}{r} 69 \\ \times\ 3 \\ \hline \end{array}$$ $$\begin{array}{r} 21 \\ \times\ 2 \\ \hline \end{array}$$

13) Multiply 2-Digits by 0, 1, 2,3

Name:	Data:	Score:

$$\begin{array}{r} 88 \\ \times\ 2 \\ \hline \end{array} \quad \begin{array}{r} 86 \\ \times\ 3 \\ \hline \end{array} \quad \begin{array}{r} 76 \\ \times\ 2 \\ \hline \end{array} \quad \begin{array}{r} 77 \\ \times\ 3 \\ \hline \end{array}$$

$$\begin{array}{r} 66 \\ \times\ 3 \\ \hline \end{array} \quad \begin{array}{r} 67 \\ \times\ 2 \\ \hline \end{array} \quad \begin{array}{r} 87 \\ \times\ 3 \\ \hline \end{array} \quad \begin{array}{r} 96 \\ \times\ 1 \\ \hline \end{array}$$

$$\begin{array}{r} 99 \\ \times\ 2 \\ \hline \end{array} \quad \begin{array}{r} 85 \\ \times\ 2 \\ \hline \end{array} \quad \begin{array}{r} 87 \\ \times\ 0 \\ \hline \end{array} \quad \begin{array}{r} 89 \\ \times\ 2 \\ \hline \end{array}$$

$$\begin{array}{r} 55 \\ \times\ 2 \\ \hline \end{array} \quad \begin{array}{r} 67 \\ \times\ 3 \\ \hline \end{array} \quad \begin{array}{r} 65 \\ \times\ 2 \\ \hline \end{array} \quad \begin{array}{r} 69 \\ \times\ 2 \\ \hline \end{array}$$

$$\begin{array}{r} 94 \\ \times\ 2 \\ \hline \end{array} \quad \begin{array}{r} 63 \\ \times\ 2 \\ \hline \end{array} \quad \begin{array}{r} 98 \\ \times\ 3 \\ \hline \end{array} \quad \begin{array}{r} 22 \\ \times\ 3 \\ \hline \end{array}$$

14) Multiply 2-Digits by 4, 5

Name:	Data:	Score:

28×4	37×5	46×4	55×5
64×5	73×4	82×5	91×4
94×4	85×4	83×5	79×5
78×5	67×5	60×4	56×4
94×4	63×5	69×4	36×5

15) Multiply 2-Digits by 4,5

Name:	Data:	Score:

$$\begin{array}{r} 88 \\ \times\ 4 \\ \hline \end{array} \quad \begin{array}{r} 86 \\ \times\ 5 \\ \hline \end{array} \quad \begin{array}{r} 76 \\ \times\ 4 \\ \hline \end{array} \quad \begin{array}{r} 77 \\ \times\ 5 \\ \hline \end{array}$$

$$\begin{array}{r} 66 \\ \times\ 5 \\ \hline \end{array} \quad \begin{array}{r} 67 \\ \times\ 4 \\ \hline \end{array} \quad \begin{array}{r} 87 \\ \times\ 5 \\ \hline \end{array} \quad \begin{array}{r} 96 \\ \times\ 4 \\ \hline \end{array}$$

$$\begin{array}{r} 99 \\ \times\ 4 \\ \hline \end{array} \quad \begin{array}{r} 85 \\ \times\ 4 \\ \hline \end{array} \quad \begin{array}{r} 87 \\ \times\ 5 \\ \hline \end{array} \quad \begin{array}{r} 89 \\ \times\ 5 \\ \hline \end{array}$$

$$\begin{array}{r} 55 \\ \times\ 5 \\ \hline \end{array} \quad \begin{array}{r} 67 \\ \times\ 5 \\ \hline \end{array} \quad \begin{array}{r} 65 \\ \times\ 4 \\ \hline \end{array} \quad \begin{array}{r} 69 \\ \times\ 4 \\ \hline \end{array}$$

$$\begin{array}{r} 94 \\ \times\ 5 \\ \hline \end{array} \quad \begin{array}{r} 63 \\ \times\ 5 \\ \hline \end{array} \quad \begin{array}{r} 98 \\ \times\ 4 \\ \hline \end{array} \quad \begin{array}{r} 22 \\ \times\ 5 \\ \hline \end{array}$$

16) Multiply 2-Digits by 5, 6

Name:	**Data:**	**Score:**

$$\begin{array}{r}25\\ \times\ 5\\ \hline\end{array}\qquad\begin{array}{r}37\\ \times\ 6\\ \hline\end{array}\qquad\begin{array}{r}46\\ \times\ 5\\ \hline\end{array}\qquad\begin{array}{r}55\\ \times\ 6\\ \hline\end{array}$$

$$\begin{array}{r}64\\ \times\ 6\\ \hline\end{array}\qquad\begin{array}{r}73\\ \times\ 5\\ \hline\end{array}\qquad\begin{array}{r}82\\ \times\ 6\\ \hline\end{array}\qquad\begin{array}{r}91\\ \times\ 5\\ \hline\end{array}$$

$$\begin{array}{r}94\\ \times\ 5\\ \hline\end{array}\qquad\begin{array}{r}85\\ \times\ 5\\ \hline\end{array}\qquad\begin{array}{r}83\\ \times\ 6\\ \hline\end{array}\qquad\begin{array}{r}79\\ \times\ 6\\ \hline\end{array}$$

$$\begin{array}{r}78\\ \times\ 6\\ \hline\end{array}\qquad\begin{array}{r}67\\ \times\ 6\\ \hline\end{array}\qquad\begin{array}{r}60\\ \times\ 5\\ \hline\end{array}\qquad\begin{array}{r}56\\ \times\ 5\\ \hline\end{array}$$

$$\begin{array}{r}97\\ \times\ 5\\ \hline\end{array}\qquad\begin{array}{r}63\\ \times\ 6\\ \hline\end{array}\qquad\begin{array}{r}69\\ \times\ 5\\ \hline\end{array}\qquad\begin{array}{r}36\\ \times\ 5\\ \hline\end{array}$$

17) Multiply 2-Digits by 5, 6

Name:	Data:	Score:

$$\begin{array}{r} 88 \\ \times\ 5 \\ \hline \end{array} \quad \begin{array}{r} 86 \\ \times\ 6 \\ \hline \end{array} \quad \begin{array}{r} 76 \\ \times\ 5 \\ \hline \end{array} \quad \begin{array}{r} 77 \\ \times\ 6 \\ \hline \end{array}$$

$$\begin{array}{r} 66 \\ \times\ 6 \\ \hline \end{array} \quad \begin{array}{r} 67 \\ \times\ 5 \\ \hline \end{array} \quad \begin{array}{r} 87 \\ \times\ 6 \\ \hline \end{array} \quad \begin{array}{r} 96 \\ \times\ 5 \\ \hline \end{array}$$

$$\begin{array}{r} 99 \\ \times\ 5 \\ \hline \end{array} \quad \begin{array}{r} 85 \\ \times\ 5 \\ \hline \end{array} \quad \begin{array}{r} 87 \\ \times\ 6 \\ \hline \end{array} \quad \begin{array}{r} 89 \\ \times\ 6 \\ \hline \end{array}$$

$$\begin{array}{r} 55 \\ \times\ 6 \\ \hline \end{array} \quad \begin{array}{r} 67 \\ \times\ 6 \\ \hline \end{array} \quad \begin{array}{r} 65 \\ \times\ 6 \\ \hline \end{array} \quad \begin{array}{r} 69 \\ \times\ 5 \\ \hline \end{array}$$

$$\begin{array}{r} 94 \\ \times\ 6 \\ \hline \end{array} \quad \begin{array}{r} 63 \\ \times\ 6 \\ \hline \end{array} \quad \begin{array}{r} 98 \\ \times\ 5 \\ \hline \end{array} \quad \begin{array}{r} 32 \\ \times\ 6 \\ \hline \end{array}$$

18) Multiply 2-Digits by 6, 7

Name:	Data:	Score:

2 5 × 6	3 7 × 7	4 6 × 6	5 5 × 7
6 4 × 7	7 3 × 6	8 2 × 7	9 1 × 6
9 4 × 6	8 5 × 6	8 3 × 7	7 9 × 7
7 8 × 7	6 7 × 7	6 0 × 6	5 6 × 6
9 7 × 6	6 3 × 7	6 9 × 6	3 6 × 6

19) Multiply 2-Digits by 6, 7

Name:	Data:	Score:

$\begin{array}{r} 88 \\ \times\ \ 6 \\ \hline \end{array}$	$\begin{array}{r} 86 \\ \times\ \ 7 \\ \hline \end{array}$	$\begin{array}{r} 76 \\ \times\ \ 6 \\ \hline \end{array}$	$\begin{array}{r} 77 \\ \times\ \ 7 \\ \hline \end{array}$
$\begin{array}{r} 66 \\ \times\ \ 7 \\ \hline \end{array}$	$\begin{array}{r} 67 \\ \times\ \ 6 \\ \hline \end{array}$	$\begin{array}{r} 87 \\ \times\ \ 7 \\ \hline \end{array}$	$\begin{array}{r} 96 \\ \times\ \ 6 \\ \hline \end{array}$
$\begin{array}{r} 99 \\ \times\ \ 6 \\ \hline \end{array}$	$\begin{array}{r} 85 \\ \times\ \ 6 \\ \hline \end{array}$	$\begin{array}{r} 87 \\ \times\ \ 7 \\ \hline \end{array}$	$\begin{array}{r} 89 \\ \times\ \ 7 \\ \hline \end{array}$
$\begin{array}{r} 55 \\ \times\ \ 7 \\ \hline \end{array}$	$\begin{array}{r} 67 \\ \times\ \ 7 \\ \hline \end{array}$	$\begin{array}{r} 65 \\ \times\ \ 7 \\ \hline \end{array}$	$\begin{array}{r} 69 \\ \times\ \ 6 \\ \hline \end{array}$
$\begin{array}{r} 94 \\ \times\ \ 7 \\ \hline \end{array}$	$\begin{array}{r} 63 \\ \times\ \ 7 \\ \hline \end{array}$	$\begin{array}{r} 98 \\ \times\ \ 6 \\ \hline \end{array}$	$\begin{array}{r} 43 \\ \times\ \ 7 \\ \hline \end{array}$

20) Multiply 2-Digits by 7, 8

Name:	Data:	Score:

$$\begin{array}{r} 25 \\ \times\ 7 \\ \hline \end{array} \quad \begin{array}{r} 37 \\ \times\ 8 \\ \hline \end{array} \quad \begin{array}{r} 46 \\ \times\ 7 \\ \hline \end{array} \quad \begin{array}{r} 55 \\ \times\ 8 \\ \hline \end{array}$$

$$\begin{array}{r} 64 \\ \times\ 8 \\ \hline \end{array} \quad \begin{array}{r} 73 \\ \times\ 7 \\ \hline \end{array} \quad \begin{array}{r} 82 \\ \times\ 7 \\ \hline \end{array} \quad \begin{array}{r} 91 \\ \times\ 7 \\ \hline \end{array}$$

$$\begin{array}{r} 94 \\ \times\ 7 \\ \hline \end{array} \quad \begin{array}{r} 85 \\ \times\ 7 \\ \hline \end{array} \quad \begin{array}{r} 83 \\ \times\ 8 \\ \hline \end{array} \quad \begin{array}{r} 79 \\ \times\ 8 \\ \hline \end{array}$$

$$\begin{array}{r} 78 \\ \times\ 8 \\ \hline \end{array} \quad \begin{array}{r} 67 \\ \times\ 8 \\ \hline \end{array} \quad \begin{array}{r} 64 \\ \times\ 7 \\ \hline \end{array} \quad \begin{array}{r} 56 \\ \times\ 7 \\ \hline \end{array}$$

$$\begin{array}{r} 97 \\ \times\ 8 \\ \hline \end{array} \quad \begin{array}{r} 63 \\ \times\ 8 \\ \hline \end{array} \quad \begin{array}{r} 69 \\ \times\ 7 \\ \hline \end{array} \quad \begin{array}{r} 36 \\ \times\ 7 \\ \hline \end{array}$$

21) Multiply 2-Digits by 7, 8

Name:	Data:	Score:

$\begin{array}{r} 88 \\ \times \; 7 \\ \hline \end{array}$ $\begin{array}{r} 86 \\ \times \; 8 \\ \hline \end{array}$ $\begin{array}{r} 76 \\ \times \; 7 \\ \hline \end{array}$ $\begin{array}{r} 77 \\ \times \; 8 \\ \hline \end{array}$

$\begin{array}{r} 66 \\ \times \; 8 \\ \hline \end{array}$ $\begin{array}{r} 67 \\ \times \; 7 \\ \hline \end{array}$ $\begin{array}{r} 87 \\ \times \; 8 \\ \hline \end{array}$ $\begin{array}{r} 96 \\ \times \; 7 \\ \hline \end{array}$

$\begin{array}{r} 99 \\ \times \; 7 \\ \hline \end{array}$ $\begin{array}{r} 85 \\ \times \; 7 \\ \hline \end{array}$ $\begin{array}{r} 87 \\ \times \; 8 \\ \hline \end{array}$ $\begin{array}{r} 89 \\ \times \; 8 \\ \hline \end{array}$

$\begin{array}{r} 55 \\ \times \; 8 \\ \hline \end{array}$ $\begin{array}{r} 67 \\ \times \; 8 \\ \hline \end{array}$ $\begin{array}{r} 65 \\ \times \; 8 \\ \hline \end{array}$ $\begin{array}{r} 69 \\ \times \; 7 \\ \hline \end{array}$

$\begin{array}{r} 94 \\ \times \; 8 \\ \hline \end{array}$ $\begin{array}{r} 63 \\ \times \; 8 \\ \hline \end{array}$ $\begin{array}{r} 98 \\ \times \; 7 \\ \hline \end{array}$ $\begin{array}{r} 43 \\ \times \; 8 \\ \hline \end{array}$

22) Multiply 2-Digits by 8,9

Name:	Data:	Score:

$\begin{array}{r} 25 \\ \times\ 8 \\ \hline \end{array}$ $\begin{array}{r} 37 \\ \times\ 8 \\ \hline \end{array}$ $\begin{array}{r} 46 \\ \times\ 8 \\ \hline \end{array}$ $\begin{array}{r} 55 \\ \times\ 9 \\ \hline \end{array}$

$\begin{array}{r} 64 \\ \times\ 9 \\ \hline \end{array}$ $\begin{array}{r} 73 \\ \times\ 8 \\ \hline \end{array}$ $\begin{array}{r} 82 \\ \times\ 8 \\ \hline \end{array}$ $\begin{array}{r} 95 \\ \times\ 8 \\ \hline \end{array}$

$\begin{array}{r} 94 \\ \times\ 8 \\ \hline \end{array}$ $\begin{array}{r} 85 \\ \times\ 8 \\ \hline \end{array}$ $\begin{array}{r} 83 \\ \times\ 9 \\ \hline \end{array}$ $\begin{array}{r} 79 \\ \times\ 9 \\ \hline \end{array}$

$\begin{array}{r} 78 \\ \times\ 9 \\ \hline \end{array}$ $\begin{array}{r} 67 \\ \times\ 9 \\ \hline \end{array}$ $\begin{array}{r} 64 \\ \times\ 9 \\ \hline \end{array}$ $\begin{array}{r} 56 \\ \times\ 9 \\ \hline \end{array}$

$\begin{array}{r} 97 \\ \times\ 9 \\ \hline \end{array}$ $\begin{array}{r} 63 \\ \times\ 9 \\ \hline \end{array}$ $\begin{array}{r} 69 \\ \times\ 8 \\ \hline \end{array}$ $\begin{array}{r} 36 \\ \times\ 8 \\ \hline \end{array}$

23) Multiply 2-Digits by 8,9

Name:	Data:	Score:

$$\begin{array}{r} 88 \\ \times\ 8 \\ \hline \end{array} \quad \begin{array}{r} 86 \\ \times\ 9 \\ \hline \end{array} \quad \begin{array}{r} 76 \\ \times\ 8 \\ \hline \end{array} \quad \begin{array}{r} 77 \\ \times\ 9 \\ \hline \end{array}$$

$$\begin{array}{r} 66 \\ \times\ 9 \\ \hline \end{array} \quad \begin{array}{r} 67 \\ \times\ 8 \\ \hline \end{array} \quad \begin{array}{r} 87 \\ \times\ 9 \\ \hline \end{array} \quad \begin{array}{r} 96 \\ \times\ 8 \\ \hline \end{array}$$

$$\begin{array}{r} 99 \\ \times\ 8 \\ \hline \end{array} \quad \begin{array}{r} 85 \\ \times\ 8 \\ \hline \end{array} \quad \begin{array}{r} 87 \\ \times\ 9 \\ \hline \end{array} \quad \begin{array}{r} 89 \\ \times\ 9 \\ \hline \end{array}$$

$$\begin{array}{r} 55 \\ \times\ 9 \\ \hline \end{array} \quad \begin{array}{r} 67 \\ \times\ 9 \\ \hline \end{array} \quad \begin{array}{r} 65 \\ \times\ 9 \\ \hline \end{array} \quad \begin{array}{r} 69 \\ \times\ 8 \\ \hline \end{array}$$

$$\begin{array}{r} 94 \\ \times\ 9 \\ \hline \end{array} \quad \begin{array}{r} 63 \\ \times\ 9 \\ \hline \end{array} \quad \begin{array}{r} 98 \\ \times\ 8 \\ \hline \end{array} \quad \begin{array}{r} 43 \\ \times\ 9 \\ \hline \end{array}$$

24) Multiply 2-Digits by 1-Digit

Name:	Data:	Score:

$$\begin{array}{r} 20 \\ \times\ 3 \\ \hline \end{array} \quad \begin{array}{r} 30 \\ \times\ 4 \\ \hline \end{array} \quad \begin{array}{r} 40 \\ \times\ 5 \\ \hline \end{array} \quad \begin{array}{r} 50 \\ \times\ 2 \\ \hline \end{array}$$

$$\begin{array}{r} 60 \\ \times\ 4 \\ \hline \end{array} \quad \begin{array}{r} 70 \\ \times\ 5 \\ \hline \end{array} \quad \begin{array}{r} 80 \\ \times\ 6 \\ \hline \end{array} \quad \begin{array}{r} 90 \\ \times\ 6 \\ \hline \end{array}$$

$$\begin{array}{r} 90 \\ \times\ 7 \\ \hline \end{array} \quad \begin{array}{r} 80 \\ \times\ 7 \\ \hline \end{array} \quad \begin{array}{r} 80 \\ \times\ 9 \\ \hline \end{array} \quad \begin{array}{r} 70 \\ \times\ 9 \\ \hline \end{array}$$

$$\begin{array}{r} 70 \\ \times\ 8 \\ \hline \end{array} \quad \begin{array}{r} 60 \\ \times\ 5 \\ \hline \end{array} \quad \begin{array}{r} 50 \\ \times\ 3 \\ \hline \end{array} \quad \begin{array}{r} 56 \\ \times\ 5 \\ \hline \end{array}$$

$$\begin{array}{r} 40 \\ \times\ 6 \\ \hline \end{array} \quad \begin{array}{r} 60 \\ \times\ 8 \\ \hline \end{array} \quad \begin{array}{r} 60 \\ \times\ 7 \\ \hline \end{array} \quad \begin{array}{r} 30 \\ \times\ 9 \\ \hline \end{array}$$

25) Multiply 2-Digits by 1-Digit

Name:	Data:	Score:

$$\begin{array}{r} 71 \\ \times\ 3 \\ \hline \end{array} \quad \begin{array}{r} 35 \\ \times\ 4 \\ \hline \end{array} \quad \begin{array}{r} 46 \\ \times\ 5 \\ \hline \end{array} \quad \begin{array}{r} 53 \\ \times\ 2 \\ \hline \end{array}$$

$$\begin{array}{r} 62 \\ \times\ 4 \\ \hline \end{array} \quad \begin{array}{r} 74 \\ \times\ 5 \\ \hline \end{array} \quad \begin{array}{r} 85 \\ \times\ 6 \\ \hline \end{array} \quad \begin{array}{r} 93 \\ \times\ 6 \\ \hline \end{array}$$

$$\begin{array}{r} 96 \\ \times\ 7 \\ \hline \end{array} \quad \begin{array}{r} 82 \\ \times\ 7 \\ \hline \end{array} \quad \begin{array}{r} 87 \\ \times\ 9 \\ \hline \end{array} \quad \begin{array}{r} 78 \\ \times\ 9 \\ \hline \end{array}$$

$$\begin{array}{r} 73 \\ \times\ 8 \\ \hline \end{array} \quad \begin{array}{r} 67 \\ \times\ 5 \\ \hline \end{array} \quad \begin{array}{r} 50 \\ \times\ 3 \\ \hline \end{array} \quad \begin{array}{r} 56 \\ \times\ 5 \\ \hline \end{array}$$

$$\begin{array}{r} 47 \\ \times\ 6 \\ \hline \end{array} \quad \begin{array}{r} 63 \\ \times\ 8 \\ \hline \end{array} \quad \begin{array}{r} 69 \\ \times\ 2 \\ \hline \end{array} \quad \begin{array}{r} 36 \\ \times\ 9 \\ \hline \end{array}$$

26) Multiply 2-Digits by 1-Digit

Name:	Data:	Score:

$$\begin{array}{r} 45 \\ \times\ 3 \\ \hline \end{array} \quad \begin{array}{r} 37 \\ \times\ 4 \\ \hline \end{array} \quad \begin{array}{r} 49 \\ \times\ 5 \\ \hline \end{array} \quad \begin{array}{r} 53 \\ \times\ 2 \\ \hline \end{array}$$

$$\begin{array}{r} 62 \\ \times\ 4 \\ \hline \end{array} \quad \begin{array}{r} 76 \\ \times\ 5 \\ \hline \end{array} \quad \begin{array}{r} 83 \\ \times\ 6 \\ \hline \end{array} \quad \begin{array}{r} 91 \\ \times\ 6 \\ \hline \end{array}$$

$$\begin{array}{r} 93 \\ \times\ 7 \\ \hline \end{array} \quad \begin{array}{r} 84 \\ \times\ 7 \\ \hline \end{array} \quad \begin{array}{r} 83 \\ \times\ 9 \\ \hline \end{array} \quad \begin{array}{r} 71 \\ \times\ 9 \\ \hline \end{array}$$

$$\begin{array}{r} 43 \\ \times\ 8 \\ \hline \end{array} \quad \begin{array}{r} 53 \\ \times\ 5 \\ \hline \end{array} \quad \begin{array}{r} 56 \\ \times\ 3 \\ \hline \end{array} \quad \begin{array}{r} 57 \\ \times\ 5 \\ \hline \end{array}$$

$$\begin{array}{r} 41 \\ \times\ 6 \\ \hline \end{array} \quad \begin{array}{r} 69 \\ \times\ 8 \\ \hline \end{array} \quad \begin{array}{r} 79 \\ \times\ 7 \\ \hline \end{array} \quad \begin{array}{r} 39 \\ \times\ 8 \\ \hline \end{array}$$

27) Multiply 2-Digits by 1-Digit

Name:	**Data:**	**Score:**

$$\begin{array}{r} 73 \\ \times\ 3 \\ \hline \end{array} \qquad \begin{array}{r} 75 \\ \times\ 4 \\ \hline \end{array} \qquad \begin{array}{r} 47 \\ \times\ 5 \\ \hline \end{array} \qquad \begin{array}{r} 58 \\ \times\ 2 \\ \hline \end{array}$$

$$\begin{array}{r} 65 \\ \times\ 4 \\ \hline \end{array} \qquad \begin{array}{r} 72 \\ \times\ 6 \\ \hline \end{array} \qquad \begin{array}{r} 83 \\ \times\ 7 \\ \hline \end{array} \qquad \begin{array}{r} 93 \\ \times\ 5 \\ \hline \end{array}$$

$$\begin{array}{r} 96 \\ \times\ 3 \\ \hline \end{array} \qquad \begin{array}{r} 82 \\ \times\ 2 \\ \hline \end{array} \qquad \begin{array}{r} 86 \\ \times\ 5 \\ \hline \end{array} \qquad \begin{array}{r} 76 \\ \times\ 7 \\ \hline \end{array}$$

$$\begin{array}{r} 73 \\ \times\ 4 \\ \hline \end{array} \qquad \begin{array}{r} 67 \\ \times\ 8 \\ \hline \end{array} \qquad \begin{array}{r} 50 \\ \times\ 8 \\ \hline \end{array} \qquad \begin{array}{r} 46 \\ \times\ 5 \\ \hline \end{array}$$

$$\begin{array}{r} 45 \\ \times\ 6 \\ \hline \end{array} \qquad \begin{array}{r} 63 \\ \times\ 8 \\ \hline \end{array} \qquad \begin{array}{r} 65 \\ \times\ 9 \\ \hline \end{array} \qquad \begin{array}{r} 86 \\ \times\ 8 \\ \hline \end{array}$$

28) Multiply 2-Digits by 1-Digit

Name:	Data:	Score:

15×3	27×4	39×5	43×2
52×4	66×5	73×6	81×6
83×9	94×9	23×7	31×7
53×8	43×5	66×3	77×5
81×6	79×8	95×7	35×8

29) Multiply 2-Digits by 1-Digit

Name:	Data:	Score:

$$\begin{array}{r} 33 \\ \times\ 3 \\ \hline \end{array} \quad \begin{array}{r} 45 \\ \times\ 4 \\ \hline \end{array} \quad \begin{array}{r} 57 \\ \times\ 5 \\ \hline \end{array} \quad \begin{array}{r} 68 \\ \times\ 2 \\ \hline \end{array}$$

$$\begin{array}{r} 75 \\ \times\ 4 \\ \hline \end{array} \quad \begin{array}{r} 82 \\ \times\ 6 \\ \hline \end{array} \quad \begin{array}{r} 53 \\ \times\ 7 \\ \hline \end{array} \quad \begin{array}{r} 63 \\ \times\ 5 \\ \hline \end{array}$$

$$\begin{array}{r} 76 \\ \times\ 3 \\ \hline \end{array} \quad \begin{array}{r} 82 \\ \times\ 2 \\ \hline \end{array} \quad \begin{array}{r} 86 \\ \times\ 5 \\ \hline \end{array} \quad \begin{array}{r} 96 \\ \times\ 7 \\ \hline \end{array}$$

$$\begin{array}{r} 43 \\ \times\ 4 \\ \hline \end{array} \quad \begin{array}{r} 57 \\ \times\ 8 \\ \hline \end{array} \quad \begin{array}{r} 66 \\ \times\ 8 \\ \hline \end{array} \quad \begin{array}{r} 76 \\ \times\ 5 \\ \hline \end{array}$$

$$\begin{array}{r} 85 \\ \times\ 6 \\ \hline \end{array} \quad \begin{array}{r} 73 \\ \times\ 8 \\ \hline \end{array} \quad \begin{array}{r} 65 \\ \times\ 7 \\ \hline \end{array} \quad \begin{array}{r} 59 \\ \times\ 8 \\ \hline \end{array}$$

30) Multiply 2-Digits by 1-Digit

Name:	Data:	Score:

$\begin{array}{r} 25 \\ \times\ 3 \\ \hline \end{array}$ $\begin{array}{r} 37 \\ \times\ 4 \\ \hline \end{array}$ $\begin{array}{r} 49 \\ \times\ 5 \\ \hline \end{array}$ $\begin{array}{r} 53 \\ \times\ 2 \\ \hline \end{array}$

$\begin{array}{r} 62 \\ \times\ 4 \\ \hline \end{array}$ $\begin{array}{r} 76 \\ \times\ 5 \\ \hline \end{array}$ $\begin{array}{r} 43 \\ \times\ 6 \\ \hline \end{array}$ $\begin{array}{r} 51 \\ \times\ 6 \\ \hline \end{array}$

$\begin{array}{r} 73 \\ \times\ 7 \\ \hline \end{array}$ $\begin{array}{r} 64 \\ \times\ 7 \\ \hline \end{array}$ $\begin{array}{r} 83 \\ \times\ 9 \\ \hline \end{array}$ $\begin{array}{r} 71 \\ \times\ 9 \\ \hline \end{array}$

$\begin{array}{r} 63 \\ \times\ 5 \\ \hline \end{array}$ $\begin{array}{r} 53 \\ \times\ 6 \\ \hline \end{array}$ $\begin{array}{r} 76 \\ \times\ 4 \\ \hline \end{array}$ $\begin{array}{r} 87 \\ \times\ 2 \\ \hline \end{array}$

$\begin{array}{r} 91 \\ \times\ 5 \\ \hline \end{array}$ $\begin{array}{r} 89 \\ \times\ 4 \\ \hline \end{array}$ $\begin{array}{r} 96 \\ \times\ 3 \\ \hline \end{array}$ $\begin{array}{r} 45 \\ \times\ 2 \\ \hline \end{array}$

31) Multiply 2-Digits by 1-Digit

Name:	Data:	Score:

$$\begin{array}{r} 13 \\ \times\ 3 \\ \hline \end{array} \qquad \begin{array}{r} 25 \\ \times\ 4 \\ \hline \end{array} \qquad \begin{array}{r} 37 \\ \times\ 5 \\ \hline \end{array} \qquad \begin{array}{r} 48 \\ \times\ 2 \\ \hline \end{array}$$

$$\begin{array}{r} 55 \\ \times\ 4 \\ \hline \end{array} \qquad \begin{array}{r} 62 \\ \times\ 6 \\ \hline \end{array} \qquad \begin{array}{r} 73 \\ \times\ 7 \\ \hline \end{array} \qquad \begin{array}{r} 83 \\ \times\ 5 \\ \hline \end{array}$$

$$\begin{array}{r} 66 \\ \times\ 3 \\ \hline \end{array} \qquad \begin{array}{r} 72 \\ \times\ 2 \\ \hline \end{array} \qquad \begin{array}{r} 56 \\ \times\ 5 \\ \hline \end{array} \qquad \begin{array}{r} 56 \\ \times\ 7 \\ \hline \end{array}$$

$$\begin{array}{r} 93 \\ \times\ 4 \\ \hline \end{array} \qquad \begin{array}{r} 87 \\ \times\ 8 \\ \hline \end{array} \qquad \begin{array}{r} 56 \\ \times\ 8 \\ \hline \end{array} \qquad \begin{array}{r} 36 \\ \times\ 5 \\ \hline \end{array}$$

$$\begin{array}{r} 35 \\ \times\ 6 \\ \hline \end{array} \qquad \begin{array}{r} 23 \\ \times\ 8 \\ \hline \end{array} \qquad \begin{array}{r} 75 \\ \times\ 9 \\ \hline \end{array} \qquad \begin{array}{r} 89 \\ \times\ 8 \\ \hline \end{array}$$

32) Multiply 2-Digits by 1-Digit

Name:	Data:	Score:

$45 \times 7 =$ $65 \times 7 =$ $74 \times 8 =$ $85 \times 8 =$

$$\begin{array}{r} 64 \\ \times\ 4 \\ \hline \end{array} \quad \begin{array}{r} 53 \\ \times\ 9 \\ \hline \end{array} \quad \begin{array}{r} 79 \\ \times\ 4 \\ \hline \end{array} \quad \begin{array}{r} 97 \\ \times\ 6 \\ \hline \end{array}$$

$$\begin{array}{r} 74 \\ \times\ 5 \\ \hline \end{array} \quad \begin{array}{r} 68 \\ \times\ 6 \\ \hline \end{array} \quad \begin{array}{r} 86 \\ \times\ 4 \\ \hline \end{array} \quad \begin{array}{r} 35 \\ \times\ 7 \\ \hline \end{array}$$

$$\begin{array}{r} 73 \\ \times\ 4 \\ \hline \end{array} \quad \begin{array}{r} 37 \\ \times\ 6 \\ \hline \end{array} \quad \begin{array}{r} 76 \\ \times\ 8 \\ \hline \end{array} \quad \begin{array}{r} 87 \\ \times\ 5 \\ \hline \end{array}$$

$52 \times 7 =$ $72 \times 7 =$ $28 \times 6 =$ $95 \times 7 =$

33) Multiply 2-Digits by 1-Digit

Name:	Data:	Score:

$$\begin{array}{r} 43 \\ \times\ 3 \\ \hline \end{array} \qquad \begin{array}{r} 44 \\ \times\ 4 \\ \hline \end{array} \qquad \begin{array}{r} 33 \\ \times\ 5 \\ \hline \end{array} \qquad \begin{array}{r} 55 \\ \times\ 5 \\ \hline \end{array}$$

$$\begin{array}{r} 55 \\ \times\ 7 \\ \hline \end{array} \qquad \begin{array}{r} 66 \\ \times\ 6 \\ \hline \end{array} \qquad \begin{array}{r} 77 \\ \times\ 7 \\ \hline \end{array} \qquad \begin{array}{r} 83 \\ \times\ 5 \\ \hline \end{array}$$

$$\begin{array}{r} 66 \\ \times\ 8 \\ \hline \end{array} \qquad \begin{array}{r} 77 \\ \times\ 8 \\ \hline \end{array} \qquad \begin{array}{r} 57 \\ \times\ 5 \\ \hline \end{array} \qquad \begin{array}{r} 57 \\ \times\ 7 \\ \hline \end{array}$$

$$\begin{array}{r} 99 \\ \times\ 4 \\ \hline \end{array} \qquad \begin{array}{r} 88 \\ \times\ 8 \\ \hline \end{array} \qquad \begin{array}{r} 86 \\ \times\ 8 \\ \hline \end{array} \qquad \begin{array}{r} 36 \\ \times\ 5 \\ \hline \end{array}$$

$55 \times 7 =$ $\qquad$ $77 \times 9 =$ $\qquad$ $88 \times 6 =$ $\qquad$ $99 \times 7 =$

34) Multiply 3-Digits by 1-Digit

Name:	Data:	Score:

$$\begin{array}{r} 124 \\ \times \quad 2 \\ \hline \end{array} \qquad \begin{array}{r} 127 \\ \times \quad 3 \\ \hline \end{array} \qquad \begin{array}{r} 104 \\ \times \quad 4 \\ \hline \end{array} \qquad \begin{array}{r} 224 \\ \times \quad 5 \\ \hline \end{array}$$

$$\begin{array}{r} 204 \\ \times \quad 6 \\ \hline \end{array} \qquad \begin{array}{r} 154 \\ \times \quad 7 \\ \hline \end{array} \qquad \begin{array}{r} 127 \\ \times \quad 8 \\ \hline \end{array} \qquad \begin{array}{r} 205 \\ \times \quad 9 \\ \hline \end{array}$$

$$\begin{array}{r} 220 \\ \times \quad 7 \\ \hline \end{array} \qquad \begin{array}{r} 118 \\ \times \quad 5 \\ \hline \end{array} \qquad \begin{array}{r} 129 \\ \times \quad 4 \\ \hline \end{array} \qquad \begin{array}{r} 154 \\ \times \quad 5 \\ \hline \end{array}$$

$$\begin{array}{r} 264 \\ \times \quad 6 \\ \hline \end{array} \qquad \begin{array}{r} 175 \\ \times \quad 7 \\ \hline \end{array} \qquad \begin{array}{r} 157 \\ \times \quad 8 \\ \hline \end{array} \qquad \begin{array}{r} 284 \\ \times \quad 9 \\ \hline \end{array}$$

$$\begin{array}{r} 256 \\ \times \quad 9 \\ \hline \end{array} \qquad \begin{array}{r} 265 \\ \times \quad 8 \\ \hline \end{array} \qquad \begin{array}{r} 190 \\ \times \quad 7 \\ \hline \end{array} \qquad \begin{array}{r} 109 \\ \times \quad 6 \\ \hline \end{array}$$

35) Multiply 3-Digits by 1-Digit

Name:	Data:	Score:

$$\begin{array}{r} 324 \\ \times \quad 2 \\ \hline \end{array} \qquad \begin{array}{r} 327 \\ \times \quad 3 \\ \hline \end{array} \qquad \begin{array}{r} 334 \\ \times \quad 4 \\ \hline \end{array} \qquad \begin{array}{r} 324 \\ \times \quad 5 \\ \hline \end{array}$$

$$\begin{array}{r} 404 \\ \times \quad 6 \\ \hline \end{array} \qquad \begin{array}{r} 454 \\ \times \quad 7 \\ \hline \end{array} \qquad \begin{array}{r} 427 \\ \times \quad 8 \\ \hline \end{array} \qquad \begin{array}{r} 405 \\ \times \quad 9 \\ \hline \end{array}$$

$$\begin{array}{r} 321 \\ \times \quad 7 \\ \hline \end{array} \qquad \begin{array}{r} 412 \\ \times \quad 5 \\ \hline \end{array} \qquad \begin{array}{r} 303 \\ \times \quad 4 \\ \hline \end{array} \qquad \begin{array}{r} 456 \\ \times \quad 5 \\ \hline \end{array}$$

$$\begin{array}{r} 368 \\ \times \quad 6 \\ \hline \end{array} \qquad \begin{array}{r} 277 \\ \times \quad 7 \\ \hline \end{array} \qquad \begin{array}{r} 406 \\ \times \quad 8 \\ \hline \end{array} \qquad \begin{array}{r} 385 \\ \times \quad 9 \\ \hline \end{array}$$

$$\begin{array}{r} 351 \\ \times \quad 9 \\ \hline \end{array} \qquad \begin{array}{r} 462 \\ \times \quad 8 \\ \hline \end{array} \qquad \begin{array}{r} 293 \\ \times \quad 7 \\ \hline \end{array} \qquad \begin{array}{r} 404 \\ \times \quad 6 \\ \hline \end{array}$$

36) Multiply 3-Digits by 1-Digit

Name:	Data:	Score:

$$\begin{array}{r} 524 \\ \times \quad 2 \\ \hline \end{array} \qquad \begin{array}{r} 527 \\ \times \quad 3 \\ \hline \end{array} \qquad \begin{array}{r} 504 \\ \times \quad 4 \\ \hline \end{array} \qquad \begin{array}{r} 524 \\ \times \quad 5 \\ \hline \end{array}$$

$$\begin{array}{r} 254 \\ \times \quad 6 \\ \hline \end{array} \qquad \begin{array}{r} 454 \\ \times \quad 7 \\ \hline \end{array} \qquad \begin{array}{r} 527 \\ \times \quad 8 \\ \hline \end{array} \qquad \begin{array}{r} 505 \\ \times \quad 9 \\ \hline \end{array}$$

$$\begin{array}{r} 520 \\ \times \quad 7 \\ \hline \end{array} \qquad \begin{array}{r} 318 \\ \times \quad 5 \\ \hline \end{array} \qquad \begin{array}{r} 529 \\ \times \quad 4 \\ \hline \end{array} \qquad \begin{array}{r} 454 \\ \times \quad 5 \\ \hline \end{array}$$

$$\begin{array}{r} 364 \\ \times \quad 6 \\ \hline \end{array} \qquad \begin{array}{r} 475 \\ \times \quad 7 \\ \hline \end{array} \qquad \begin{array}{r} 457 \\ \times \quad 8 \\ \hline \end{array} \qquad \begin{array}{r} 584 \\ \times \quad 9 \\ \hline \end{array}$$

$$\begin{array}{r} 356 \\ \times \quad 9 \\ \hline \end{array} \qquad \begin{array}{r} 465 \\ \times \quad 8 \\ \hline \end{array} \qquad \begin{array}{r} 590 \\ \times \quad 7 \\ \hline \end{array} \qquad \begin{array}{r} 509 \\ \times \quad 6 \\ \hline \end{array}$$

37) Multiply 3-Digits by 1-Digit

Name:	Data:	Score:

$$\begin{array}{r} 624 \\ \times \quad 2 \\ \hline \end{array} \qquad \begin{array}{r} 627 \\ \times \quad 3 \\ \hline \end{array} \qquad \begin{array}{r} 634 \\ \times \quad 4 \\ \hline \end{array} \qquad \begin{array}{r} 624 \\ \times \quad 5 \\ \hline \end{array}$$

$$\begin{array}{r} 504 \\ \times \quad 6 \\ \hline \end{array} \qquad \begin{array}{r} 654 \\ \times \quad 7 \\ \hline \end{array} \qquad \begin{array}{r} 627 \\ \times \quad 8 \\ \hline \end{array} \qquad \begin{array}{r} 605 \\ \times \quad 9 \\ \hline \end{array}$$

$$\begin{array}{r} 421 \\ \times \quad 4 \\ \hline \end{array} \qquad \begin{array}{r} 512 \\ \times \quad 3 \\ \hline \end{array} \qquad \begin{array}{r} 603 \\ \times \quad 2 \\ \hline \end{array} \qquad \begin{array}{r} 356 \\ \times \quad 5 \\ \hline \end{array}$$

$$\begin{array}{r} 568 \\ \times \quad 6 \\ \hline \end{array} \qquad \begin{array}{r} 677 \\ \times \quad 7 \\ \hline \end{array} \qquad \begin{array}{r} 506 \\ \times \quad 8 \\ \hline \end{array} \qquad \begin{array}{r} 685 \\ \times \quad 9 \\ \hline \end{array}$$

$$\begin{array}{r} 651 \\ \times \quad 9 \\ \hline \end{array} \qquad \begin{array}{r} 562 \\ \times \quad 8 \\ \hline \end{array} \qquad \begin{array}{r} 693 \\ \times \quad 7 \\ \hline \end{array} \qquad \begin{array}{r} 604 \\ \times \quad 6 \\ \hline \end{array}$$

38) Multiply 3-Digits by 1-Digit

Name:	Data:	Score:

$$\begin{array}{r} 734 \\ \times \ \ 2 \\ \hline \end{array} \qquad \begin{array}{r} 437 \\ \times \ \ 3 \\ \hline \end{array} \qquad \begin{array}{r} 714 \\ \times \ \ 4 \\ \hline \end{array} \qquad \begin{array}{r} 734 \\ \times \ \ 5 \\ \hline \end{array}$$

$$\begin{array}{r} 764 \\ \times \ \ 6 \\ \hline \end{array} \qquad \begin{array}{r} 764 \\ \times \ \ 7 \\ \hline \end{array} \qquad \begin{array}{r} 737 \\ \times \ \ 8 \\ \hline \end{array} \qquad \begin{array}{r} 715 \\ \times \ \ 9 \\ \hline \end{array}$$

$$\begin{array}{r} 730 \\ \times \ \ 7 \\ \hline \end{array} \qquad \begin{array}{r} 628 \\ \times \ \ 5 \\ \hline \end{array} \qquad \begin{array}{r} 709 \\ \times \ \ 4 \\ \hline \end{array} \qquad \begin{array}{r} 704 \\ \times \ \ 5 \\ \hline \end{array}$$

$$\begin{array}{r} 704 \\ \times \ \ 3 \\ \hline \end{array} \qquad \begin{array}{r} 605 \\ \times \ \ 4 \\ \hline \end{array} \qquad \begin{array}{r} 507 \\ \times \ \ 5 \\ \hline \end{array} \qquad \begin{array}{r} 704 \\ \times \ \ 9 \\ \hline \end{array}$$

$$\begin{array}{r} 706 \\ \times \ \ 9 \\ \hline \end{array} \qquad \begin{array}{r} 705 \\ \times \ \ 8 \\ \hline \end{array} \qquad \begin{array}{r} 706 \\ \times \ \ 7 \\ \hline \end{array} \qquad \begin{array}{r} 609 \\ \times \ \ 6 \\ \hline \end{array}$$

39) Multiply 3-Digits by 1-Digit

Name:	Data:	Score:

$$\begin{array}{r} 823 \\ \times \quad 2 \\ \hline \end{array} \qquad \begin{array}{r} 823 \\ \times \quad 3 \\ \hline \end{array} \qquad \begin{array}{r} 834 \\ \times \quad 4 \\ \hline \end{array} \qquad \begin{array}{r} 823 \\ \times \quad 5 \\ \hline \end{array}$$

$$\begin{array}{r} 805 \\ \times \quad 6 \\ \hline \end{array} \qquad \begin{array}{r} 755 \\ \times \quad 7 \\ \hline \end{array} \qquad \begin{array}{r} 825 \\ \times \quad 8 \\ \hline \end{array} \qquad \begin{array}{r} 705 \\ \times \quad 9 \\ \hline \end{array}$$

$$\begin{array}{r} 824 \\ \times \quad 4 \\ \hline \end{array} \qquad \begin{array}{r} 814 \\ \times \quad 3 \\ \hline \end{array} \qquad \begin{array}{r} 704 \\ \times \quad 2 \\ \hline \end{array} \qquad \begin{array}{r} 854 \\ \times \quad 5 \\ \hline \end{array}$$

$$\begin{array}{r} 867 \\ \times \quad 6 \\ \hline \end{array} \qquad \begin{array}{r} 878 \\ \times \quad 7 \\ \hline \end{array} \qquad \begin{array}{r} 807 \\ \times \quad 8 \\ \hline \end{array} \qquad \begin{array}{r} 895 \\ \times \quad 9 \\ \hline \end{array}$$

$$\begin{array}{r} 852 \\ \times \quad 9 \\ \hline \end{array} \qquad \begin{array}{r} 862 \\ \times \quad 8 \\ \hline \end{array} \qquad \begin{array}{r} 894 \\ \times \quad 7 \\ \hline \end{array} \qquad \begin{array}{r} 805 \\ \times \quad 6 \\ \hline \end{array}$$

40) Multiply 3-Digits by 1-Digit

Name:	Data:	Score:

$$\begin{array}{r} 936 \\ \times\ 2 \\ \hline \end{array} \quad \begin{array}{r} 439 \\ \times\ 3 \\ \hline \end{array} \quad \begin{array}{r} 514 \\ \times\ 4 \\ \hline \end{array} \quad \begin{array}{r} 937 \\ \times\ 5 \\ \hline \end{array}$$

$$\begin{array}{r} 967 \\ \times\ 6 \\ \hline \end{array} \quad \begin{array}{r} 468 \\ \times\ 7 \\ \hline \end{array} \quad \begin{array}{r} 535 \\ \times\ 8 \\ \hline \end{array} \quad \begin{array}{r} 915 \\ \times\ 9 \\ \hline \end{array}$$

$$\begin{array}{r} 908 \\ \times\ 7 \\ \hline \end{array} \quad \begin{array}{r} 407 \\ \times\ 5 \\ \hline \end{array} \quad \begin{array}{r} 506 \\ \times\ 4 \\ \hline \end{array} \quad \begin{array}{r} 906 \\ \times\ 5 \\ \hline \end{array}$$

$$\begin{array}{r} 904 \\ \times\ 3 \\ \hline \end{array} \quad \begin{array}{r} 406 \\ \times\ 4 \\ \hline \end{array} \quad \begin{array}{r} 508 \\ \times\ 6 \\ \hline \end{array} \quad \begin{array}{r} 938 \\ \times\ 9 \\ \hline \end{array}$$

$$\begin{array}{r} 736 \\ \times\ 2 \\ \hline \end{array} \quad \begin{array}{r} 745 \\ \times\ 8 \\ \hline \end{array} \quad \begin{array}{r} 756 \\ \times\ 2 \\ \hline \end{array} \quad \begin{array}{r} 679 \\ \times\ 6 \\ \hline \end{array}$$

41) Multiply 3-Digits by 1-Digit

Name:	Data:	Score:

924×2	724×3	635×4	724×5
306×6	256×7	126×8	406×9
526×4	615×3	405×2	756×5
268×6	379×7	408×8	596×9
653×9	763×8	293×7	304×6

42) Multiply 3-Digits by 1-Digit

Name:	Data:	Score:

$$\begin{array}{r} 1034 \\ \times \quad 2 \\ \hline \end{array} \qquad \begin{array}{r} 2204 \\ \times \quad 3 \\ \hline \end{array} \qquad \begin{array}{r} 3034 \\ \times \quad 4 \\ \hline \end{array} \qquad \begin{array}{r} 4254 \\ \times \quad 5 \\ \hline \end{array}$$

$$\begin{array}{r} 5204 \\ \times \quad 6 \\ \hline \end{array} \qquad \begin{array}{r} 6254 \\ \times \quad 7 \\ \hline \end{array} \qquad \begin{array}{r} 2213 \\ \times \quad 8 \\ \hline \end{array} \qquad \begin{array}{r} 1272 \\ \times \quad 9 \\ \hline \end{array}$$

$$\begin{array}{r} 8037 \\ \times \quad 2 \\ \hline \end{array} \qquad \begin{array}{r} 6276 \\ \times \quad 3 \\ \hline \end{array} \qquad \begin{array}{r} 7534 \\ \times \quad 4 \\ \hline \end{array} \qquad \begin{array}{r} 8935 \\ \times \quad 5 \\ \hline \end{array}$$

$$\begin{array}{r} 3204 \\ \times \quad 6 \\ \hline \end{array} \qquad \begin{array}{r} 3839 \\ \times \quad 7 \\ \hline \end{array} \qquad \begin{array}{r} 1248 \\ \times \quad 8 \\ \hline \end{array} \qquad \begin{array}{r} 4237 \\ \times \quad 9 \\ \hline \end{array}$$

$$\begin{array}{r} 6035 \\ \times \quad 4 \\ \hline \end{array} \qquad \begin{array}{r} 5237 \\ \times \quad 6 \\ \hline \end{array} \qquad \begin{array}{r} 1034 \\ \times \quad 8 \\ \hline \end{array} \qquad \begin{array}{r} 9254 \\ \times \quad 2 \\ \hline \end{array}$$

43) Multiply 3-Digits by 1-Digit

Name:	Data:	Score:

$$\begin{array}{r} 2035 \\ \times\quad 2 \\ \hline \end{array}$$ $$\begin{array}{r} 3205 \\ \times\quad 3 \\ \hline \end{array}$$ $$\begin{array}{r} 4635 \\ \times\quad 4 \\ \hline \end{array}$$ $$\begin{array}{r} 5205 \\ \times\quad 5 \\ \hline \end{array}$$

$$\begin{array}{r} 6205 \\ \times\quad 6 \\ \hline \end{array}$$ $$\begin{array}{r} 7215 \\ \times\quad 7 \\ \hline \end{array}$$ $$\begin{array}{r} 8235 \\ \times\quad 8 \\ \hline \end{array}$$ $$\begin{array}{r} 9134 \\ \times\quad 9 \\ \hline \end{array}$$

$$\begin{array}{r} 1038 \\ \times\quad 2 \\ \hline \end{array}$$ $$\begin{array}{r} 2247 \\ \times\quad 3 \\ \hline \end{array}$$ $$\begin{array}{r} 3035 \\ \times\quad 4 \\ \hline \end{array}$$ $$\begin{array}{r} 4246 \\ \times\quad 5 \\ \hline \end{array}$$

$$\begin{array}{r} 5205 \\ \times\quad 6 \\ \hline \end{array}$$ $$\begin{array}{r} 6538 \\ \times\quad 7 \\ \hline \end{array}$$ $$\begin{array}{r} 7239 \\ \times\quad 8 \\ \hline \end{array}$$ $$\begin{array}{r} 8538 \\ \times\quad 9 \\ \hline \end{array}$$

$$\begin{array}{r} 9036 \\ \times\quad 4 \\ \hline \end{array}$$ $$\begin{array}{r} 1738 \\ \times\quad 6 \\ \hline \end{array}$$ $$\begin{array}{r} 2035 \\ \times\quad 8 \\ \hline \end{array}$$ $$\begin{array}{r} 3256 \\ \times\quad 2 \\ \hline \end{array}$$

Congratulations! You have completed one digit by any number of digits!

Time to relax and enjoy coloring using your choice of colors!

44) Multiply 2-Digits by 2-Digits

Name:	Data:	Score:

32×14

1	2	8
3	2	**0**
4	4	8

14×24

	5	6
2	8	**0**
3	3	8

12×10

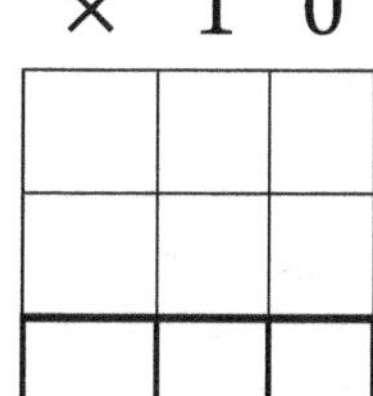

13×12

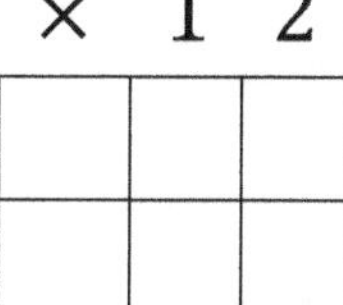

15×11

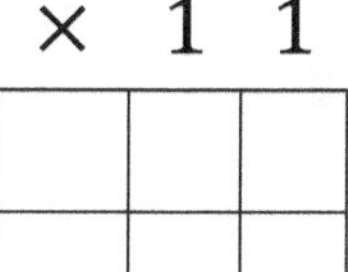

22×13

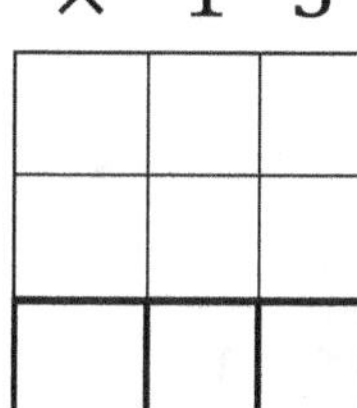

23×14

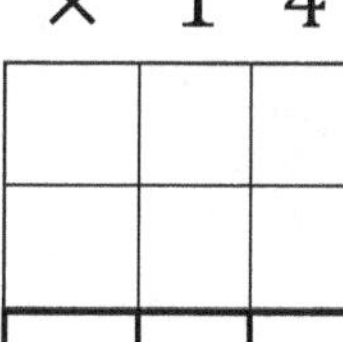

25×10

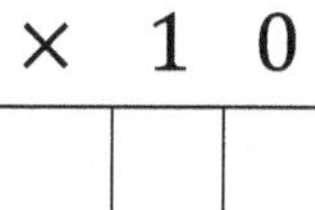

30×12

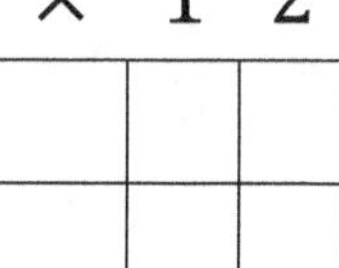

26×12

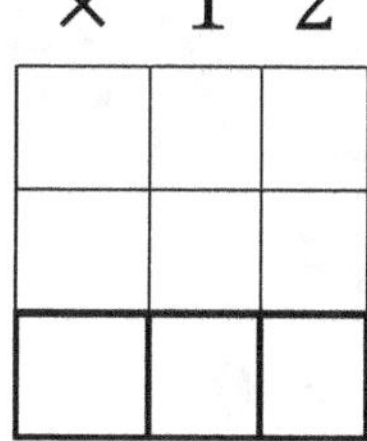

28×11

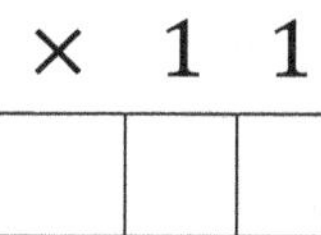

27×13

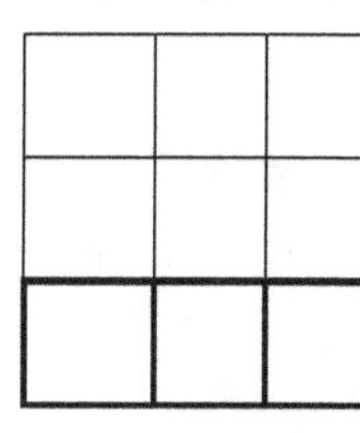

31×20

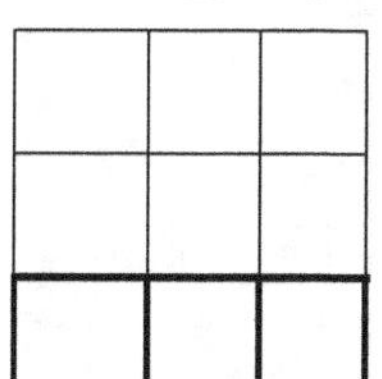

32×15

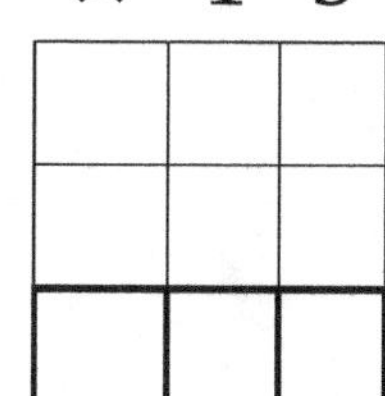

35×13

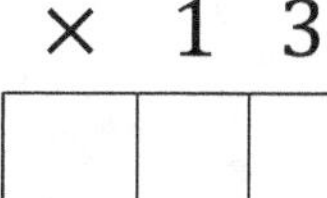

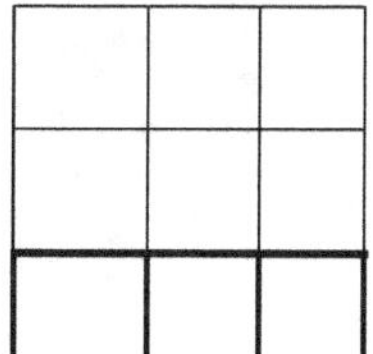

36×11

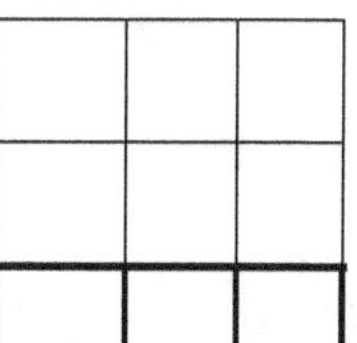

45) Multiply 2-Digits by 2-Digits

Name:	Data:	Score:

$\begin{array}{r} 28 \\ \times 10 \\ \hline \end{array}$	$\begin{array}{r} 37 \\ \times 11 \\ \hline \end{array}$	$\begin{array}{r} 46 \\ \times 12 \\ \hline \end{array}$	$\begin{array}{r} 54 \\ \times 13 \\ \hline \end{array}$
$\begin{array}{r} 14 \\ \times 13 \\ \hline \end{array}$	$\begin{array}{r} 13 \\ \times 12 \\ \hline \end{array}$	$\begin{array}{r} 42 \\ \times 13 \\ \hline \end{array}$	$\begin{array}{r} 51 \\ \times 11 \\ \hline \end{array}$
$\begin{array}{r} 54 \\ \times 12 \\ \hline \end{array}$	$\begin{array}{r} 35 \\ \times 22 \\ \hline \end{array}$	$\begin{array}{r} 33 \\ \times 20 \\ \hline \end{array}$	$\begin{array}{r} 29 \\ \times 32 \\ \hline \end{array}$
$\begin{array}{r} 45 \\ \times 12 \\ \hline \end{array}$	$\begin{array}{r} 31 \\ \times 23 \\ \hline \end{array}$	$\begin{array}{r} 20 \\ \times 32 \\ \hline \end{array}$	$\begin{array}{r} 16 \\ \times 12 \\ \hline \end{array}$

46) Multiply 2-Digits by 2-Digits

Name:	Data:	Score:

```
    4 5        6 2        5 6        3 9
  × 2 5      × 3 4      × 2 4      × 2 7
  2 2 5      2 4 8
  9 0 0    1 8 6 0
1 1 2 5    2 1 0 8
```

```
    5 6        5 6        5 6        5 6
  × 2 4      × 2 4      × 2 4      × 2 4
```

```
    5 6        5 6        5 6        5 6
  × 2 4      × 2 4      × 2 4      × 2 4
```

```
    5 6        5 6        5 6        5 6
  × 2 4      × 2 4      × 2 4      × 2 4
```

47) Multiply 2-Digits by 2-Digits

Name:	Data:	Score:

$$\begin{array}{r} 28 \\ \times 13 \\ \hline \end{array} \quad \begin{array}{r} 37 \\ \times 14 \\ \hline \end{array} \quad \begin{array}{r} 46 \\ \times 15 \\ \hline \end{array} \quad \begin{array}{r} 54 \\ \times 16 \\ \hline \end{array}$$

$$\begin{array}{r} 14 \\ \times 23 \\ \hline \end{array} \quad \begin{array}{r} 13 \\ \times 32 \\ \hline \end{array} \quad \begin{array}{r} 42 \\ \times 43 \\ \hline \end{array} \quad \begin{array}{r} 51 \\ \times 51 \\ \hline \end{array}$$

$$\begin{array}{r} 54 \\ \times 50 \\ \hline \end{array} \quad \begin{array}{r} 35 \\ \times 32 \\ \hline \end{array} \quad \begin{array}{r} 30 \\ \times 24 \\ \hline \end{array} \quad \begin{array}{r} 69 \\ \times 30 \\ \hline \end{array}$$

$$\begin{array}{r} 65 \\ \times 25 \\ \hline \end{array} \quad \begin{array}{r} 81 \\ \times 24 \\ \hline \end{array} \quad \begin{array}{r} 70 \\ \times 40 \\ \hline \end{array} \quad \begin{array}{r} 76 \\ \times 15 \\ \hline \end{array}$$

48) Multiply 2-Digits by 2-Digits

Name:	Data:	Score:

$$\begin{array}{r} 28 \\ \times 44 \\ \hline \end{array}$$

$$\begin{array}{r} 37 \\ \times 25 \\ \hline \end{array}$$

$$\begin{array}{r} 46 \\ \times 53 \\ \hline \end{array}$$

$$\begin{array}{r} 55 \\ \times 45 \\ \hline \end{array}$$

$$\begin{array}{r} 64 \\ \times 42 \\ \hline \end{array}$$

$$\begin{array}{r} 73 \\ \times 34 \\ \hline \end{array}$$

$$\begin{array}{r} 82 \\ \times 40 \\ \hline \end{array}$$

$$\begin{array}{r} 91 \\ \times 56 \\ \hline \end{array}$$

$$\begin{array}{r} 94 \\ \times 34 \\ \hline \end{array}$$

$$\begin{array}{r} 85 \\ \times 24 \\ \hline \end{array}$$

$$\begin{array}{r} 83 \\ \times 65 \\ \hline \end{array}$$

$$\begin{array}{r} 78 \\ \times 51 \\ \hline \end{array}$$

$$\begin{array}{r} 67 \\ \times 60 \\ \hline \end{array}$$

$$\begin{array}{r} 79 \\ \times 45 \\ \hline \end{array}$$

49) Multiply 2-Digits by 2-Digits

Name:	Data:	Score:

$$\begin{array}{r} 28 \\ \times 14 \\ \hline \end{array} \quad \begin{array}{r} 36 \\ \times 25 \\ \hline \end{array} \quad \begin{array}{r} 26 \\ \times 34 \\ \hline \end{array} \quad \begin{array}{r} 37 \\ \times 45 \\ \hline \end{array}$$

$$\begin{array}{r} 46 \\ \times 25 \\ \hline \end{array} \quad \begin{array}{r} 47 \\ \times 34 \\ \hline \end{array} \quad \begin{array}{r} 57 \\ \times 45 \\ \hline \end{array} \quad \begin{array}{r} 36 \\ \times 54 \\ \hline \end{array}$$

$$\begin{array}{r} 29 \\ \times 16 \\ \hline \end{array} \quad \begin{array}{r} 80 \\ \times 30 \\ \hline \end{array} \quad \begin{array}{r} 17 \\ \times 25 \\ \hline \end{array}$$

$$\begin{array}{r} 45 \\ \times 65 \\ \hline \end{array} \quad \begin{array}{r} 37 \\ \times 15 \\ \hline \end{array} \quad \begin{array}{r} 25 \\ \times 24 \\ \hline \end{array}$$

50) Multiply 2-Digits by 2-Digits

Name:	Data:	Score:

$$\begin{array}{r} 25 \\ \times 15 \\ \hline \end{array} \qquad \begin{array}{r} 37 \\ \times 26 \\ \hline \end{array} \qquad \begin{array}{r} 46 \\ \times 25 \\ \hline \end{array} \qquad \begin{array}{r} 55 \\ \times 36 \\ \hline \end{array}$$

$$\begin{array}{r} 64 \\ \times 40 \\ \hline \end{array} \qquad \begin{array}{r} 73 \\ \times 15 \\ \hline \end{array} \qquad \begin{array}{r} 82 \\ \times 30 \\ \hline \end{array} \qquad \begin{array}{r} 91 \\ \times 45 \\ \hline \end{array}$$

$$\begin{array}{r} 94 \\ \times 40 \\ \hline \end{array} \qquad \begin{array}{r} 75 \\ \times 30 \\ \hline \end{array} \qquad \begin{array}{r} 83 \\ \times 50 \\ \hline \end{array}$$

$$\begin{array}{r} 78 \\ \times 16 \\ \hline \end{array} \qquad \begin{array}{r} 67 \\ \times 26 \\ \hline \end{array} \qquad \begin{array}{r} 60 \\ \times 25 \\ \hline \end{array}$$

51) Multiply 2-Digits by 2-Digits

Name:	Data:	Score:

$$\begin{array}{r} 48 \\ \times 11 \\ \hline \end{array} \quad \begin{array}{r} 56 \\ \times 12 \\ \hline \end{array} \quad \begin{array}{r} 66 \\ \times 20 \\ \hline \end{array} \quad \begin{array}{r} 77 \\ \times 30 \\ \hline \end{array}$$

$$\begin{array}{r} 26 \\ \times 40 \\ \hline \end{array} \quad \begin{array}{r} 37 \\ \times 50 \\ \hline \end{array} \quad \begin{array}{r} 47 \\ \times 60 \\ \hline \end{array} \quad \begin{array}{r} 56 \\ \times 70 \\ \hline \end{array}$$

$$\begin{array}{r} 69 \\ \times 11 \\ \hline \end{array} \quad \begin{array}{r} 85 \\ \times 21 \\ \hline \end{array} \quad \begin{array}{r} 87 \\ \times 31 \\ \hline \end{array} \quad \begin{array}{r} 39 \\ \times 41 \\ \hline \end{array}$$

$$\begin{array}{r} 75 \\ \times 51 \\ \hline \end{array} \quad \begin{array}{r} 87 \\ \times 42 \\ \hline \end{array}$$

52) Multiply 2-Digits by 2-Digits

Name:	Data:	Score:

$\begin{array}{r} 79 \\ \times 60 \\ \hline \end{array}$ $\begin{array}{r} 60 \\ \times 41 \\ \hline \end{array}$ $\begin{array}{r} 29 \\ \times 50 \\ \hline \end{array}$ $\begin{array}{r} 75 \\ \times 52 \\ \hline \end{array}$

$\begin{array}{r} 56 \\ \times 35 \\ \hline \end{array}$ $\begin{array}{r} 69 \\ \times 24 \\ \hline \end{array}$ $\begin{array}{r} 60 \\ \times 24 \\ \hline \end{array}$ $\begin{array}{r} 98 \\ \times 35 \\ \hline \end{array}$

$\begin{array}{r} 36 \\ \times 65 \\ \hline \end{array}$ $\begin{array}{r} 36 \\ \times 35 \\ \hline \end{array}$ $\begin{array}{r} 26 \\ \times 25 \\ \hline \end{array}$

$\begin{array}{r} 56 \\ \times 27 \\ \hline \end{array}$ $\begin{array}{r} 69 \\ \times 23 \\ \hline \end{array}$ $\begin{array}{r} 32 \\ \times 46 \\ \hline \end{array}$

53) Multiply 2-Digits by 2-Digits

Name:	Data:	Score:

$$\begin{array}{r} 94 \\ \times 34 \\ \hline \end{array} \qquad \begin{array}{r} 63 \\ \times 15 \\ \hline \end{array} \qquad \begin{array}{r} 94 \\ \times 25 \\ \hline \end{array} \qquad \begin{array}{r} 63 \\ \times 23 \\ \hline \end{array}$$

$$\begin{array}{r} 28 \\ \times 45 \\ \hline \end{array} \qquad \begin{array}{r} 43 \\ \times 55 \\ \hline \end{array} \qquad \begin{array}{r} 28 \\ \times 64 \\ \hline \end{array} \qquad \begin{array}{r} 96 \\ \times 26 \\ \hline \end{array}$$

$$\begin{array}{r} 97 \\ \times 25 \\ \hline \end{array} \qquad \begin{array}{r} 63 \\ \times 35 \\ \hline \end{array} \qquad \begin{array}{r} 87 \\ \times 17 \\ \hline \end{array}$$

$$\begin{array}{r} 69 \\ \times 45 \\ \hline \end{array} \qquad \begin{array}{r} 67 \\ \times 57 \\ \hline \end{array} \qquad \begin{array}{r} 65 \\ \times 87 \\ \hline \end{array}$$

54) Multiply 2-Digits by 2-Digits

Name:	Data:	Score:

$$\begin{array}{r} 25 \\ \times 17 \\ \hline \end{array}$$

$$\begin{array}{r} 37 \\ \times 23 \\ \hline \end{array}$$

$$\begin{array}{r} 46 \\ \times 34 \\ \hline \end{array}$$

$$\begin{array}{r} 55 \\ \times 45 \\ \hline \end{array}$$

$$\begin{array}{r} 64 \\ \times 56 \\ \hline \end{array}$$

$$\begin{array}{r} 73 \\ \times 67 \\ \hline \end{array}$$

$$\begin{array}{r} 82 \\ \times 87 \\ \hline \end{array}$$

$$\begin{array}{r} 91 \\ \times 70 \\ \hline \end{array}$$

$$\begin{array}{r} 94 \\ \times 91 \\ \hline \end{array}$$

$$\begin{array}{r} 85 \\ \times 23 \\ \hline \end{array}$$

$$\begin{array}{r} 83 \\ \times 16 \\ \hline \end{array}$$

$$\begin{array}{r} 79 \\ \times 68 \\ \hline \end{array}$$

$$\begin{array}{r} 78 \\ \times 40 \\ \hline \end{array}$$

$$\begin{array}{r} 67 \\ \times 28 \\ \hline \end{array}$$

$$\begin{array}{r} 64 \\ \times 43 \\ \hline \end{array}$$

55) Multiply 2-Digits by 2-Digits

Name:	Data:	Score:

$\begin{array}{r} 84 \\ \times 12 \\ \hline \end{array}$ $\begin{array}{r} 86 \\ \times 25 \\ \hline \end{array}$ $\begin{array}{r} 76 \\ \times 37 \\ \hline \end{array}$ $\begin{array}{r} 73 \\ \times 8 \\ \hline \end{array}$

$\begin{array}{r} 66 \\ \times 50 \\ \hline \end{array}$ $\begin{array}{r} 67 \\ \times 43 \\ \hline \end{array}$ $\begin{array}{r} 87 \\ \times 28 \\ \hline \end{array}$ $\begin{array}{r} 96 \\ \times 57 \\ \hline \end{array}$

$\begin{array}{r} 90 \\ \times 17 \\ \hline \end{array}$ $\begin{array}{r} 85 \\ \times 97 \\ \hline \end{array}$ $\begin{array}{r} 87 \\ \times 49 \\ \hline \end{array}$ $\begin{array}{r} 89 \\ \times 68 \\ \hline \end{array}$

$\begin{array}{r} 55 \\ \times 56 \\ \hline \end{array}$ $\begin{array}{r} 67 \\ \times 48 \\ \hline \end{array}$ $\begin{array}{r} 65 \\ \times 38 \\ \hline \end{array}$

56) Multiply 2-Digits by 2-Digits

Name:	Data:	Score:

$\begin{array}{r} 25 \\ \times 38 \\ \hline \end{array}$	$\begin{array}{r} 37 \\ \times 25 \\ \hline \end{array}$	$\begin{array}{r} 46 \\ \times 42 \\ \hline \end{array}$	$\begin{array}{r} 55 \\ \times 32 \\ \hline \end{array}$
$\begin{array}{r} 64 \\ \times 19 \\ \hline \end{array}$	$\begin{array}{r} 73 \\ \times 28 \\ \hline \end{array}$	$\begin{array}{r} 82 \\ \times 38 \\ \hline \end{array}$	$\begin{array}{r} 95 \\ \times 50 \\ \hline \end{array}$
$\begin{array}{r} 94 \\ \times 28 \\ \hline \end{array}$	$\begin{array}{r} 85 \\ \times 60 \\ \hline \end{array}$	$\begin{array}{r} 83 \\ \times 14 \\ \hline \end{array}$	$\begin{array}{r} 79 \\ \times 62 \\ \hline \end{array}$
$\begin{array}{r} 78 \\ \times 49 \\ \hline \end{array}$	$\begin{array}{r} 67 \\ \times 62 \\ \hline \end{array}$	$\begin{array}{r} 64 \\ \times 31 \\ \hline \end{array}$	
$\begin{array}{r} 91 \\ \times 24 \\ \hline \end{array}$	$\begin{array}{r} 63 \\ \times 35 \\ \hline \end{array}$	$\begin{array}{r} 69 \\ \times 28 \\ \hline \end{array}$	

57) Multiply 2-Digits by 2-Digits

Name:	Data:	Score:

$$\begin{array}{r} 80 \\ \times 25 \\ \hline \end{array} \quad \begin{array}{r} 86 \\ \times 29 \\ \hline \end{array} \quad \begin{array}{r} 76 \\ \times 48 \\ \hline \end{array} \quad \begin{array}{r} 77 \\ \times 39 \\ \hline \end{array}$$

$$\begin{array}{r} 66 \\ \times 49 \\ \hline \end{array} \quad \begin{array}{r} 67 \\ \times 48 \\ \hline \end{array} \quad \begin{array}{r} 87 \\ \times 39 \\ \hline \end{array} \quad \begin{array}{r} 96 \\ \times 28 \\ \hline \end{array}$$

$$\begin{array}{r} 90 \\ \times 48 \\ \hline \end{array} \quad \begin{array}{r} 85 \\ \times 38 \\ \hline \end{array} \quad \begin{array}{r} 87 \\ \times 29 \\ \hline \end{array} \quad \begin{array}{r} 89 \\ \times 40 \\ \hline \end{array}$$

$$\begin{array}{r} 55 \\ \times 19 \\ \hline \end{array} \quad \begin{array}{r} 67 \\ \times 48 \\ \hline \end{array} \quad \begin{array}{r} 65 \\ \times 64 \\ \hline \end{array}$$

$$\begin{array}{r} 94 \\ \times 30 \\ \hline \end{array} \quad \begin{array}{r} 63 \\ \times 51 \\ \hline \end{array} \quad \begin{array}{r} 75 \\ \times 18 \\ \hline \end{array}$$

58) Multiply 3-Digits by 2-Digits

Name:	**Data:**	**Score:**

$$\begin{array}{r} 124 \\ \times\ 12 \\ \hline \end{array} \quad \begin{array}{r} 127 \\ \times\ 23 \\ \hline \end{array} \quad \begin{array}{r} 104 \\ \times\ 34 \\ \hline \end{array} \quad \begin{array}{r} 224 \\ \times\ 45 \\ \hline \end{array}$$

$$\begin{array}{r} 204 \\ \times\ 56 \\ \hline \end{array} \quad \begin{array}{r} 154 \\ \times\ 27 \\ \hline \end{array} \quad \begin{array}{r} 127 \\ \times\ 48 \\ \hline \end{array} \quad \begin{array}{r} 205 \\ \times\ 29 \\ \hline \end{array}$$

$$\begin{array}{r} 220 \\ \times\ 37 \\ \hline \end{array} \quad \begin{array}{r} 118 \\ \times\ 45 \\ \hline \end{array} \quad \begin{array}{r} 129 \\ \times\ 54 \\ \hline \end{array} \quad \begin{array}{r} 154 \\ \times\ 65 \\ \hline \end{array}$$

$$\begin{array}{r} 264 \\ \times\ 26 \\ \hline \end{array} \quad \begin{array}{r} 175 \\ \times\ 47 \\ \hline \end{array} \quad \begin{array}{r} 157 \\ \times\ 38 \\ \hline \end{array}$$

59) Multiply 3-Digits by 2-Digits

Name:	Data:	Score:

$$\begin{array}{r} 324 \\ \times\ 32 \\ \hline \end{array} \qquad \begin{array}{r} 327 \\ \times\ 23 \\ \hline \end{array} \qquad \begin{array}{r} 334 \\ \times\ 14 \\ \hline \end{array} \qquad \begin{array}{r} 324 \\ \times\ 35 \\ \hline \end{array}$$

$$\begin{array}{r} 404 \\ \times\ 46 \\ \hline \end{array} \qquad \begin{array}{r} 454 \\ \times\ 57 \\ \hline \end{array} \qquad \begin{array}{r} 427 \\ \times\ 48 \\ \hline \end{array} \qquad \begin{array}{r} 405 \\ \times\ 59 \\ \hline \end{array}$$

$$\begin{array}{r} 320 \\ \times\ 37 \\ \hline \end{array} \qquad \begin{array}{r} 412 \\ \times\ 25 \\ \hline \end{array} \qquad \begin{array}{r} 303 \\ \times\ 54 \\ \hline \end{array} \qquad \begin{array}{r} 450 \\ \times\ 65 \\ \hline \end{array}$$

$$\begin{array}{r} 308 \\ \times\ 36 \\ \hline \end{array} \qquad \begin{array}{r} 270 \\ \times\ 17 \\ \hline \end{array} \qquad \begin{array}{r} 406 \\ \times\ 28 \\ \hline \end{array}$$

60) Multiply 3-Digits by 2-Digits

Name:	Data:	Score:

$$\begin{array}{r} 524 \\ \times\ 62 \\ \hline \end{array} \quad \begin{array}{r} 527 \\ \times\ 53 \\ \hline \end{array} \quad \begin{array}{r} 504 \\ \times\ 24 \\ \hline \end{array} \quad \begin{array}{r} 524 \\ \times\ 35 \\ \hline \end{array}$$

$$\begin{array}{r} 254 \\ \times\ 46 \\ \hline \end{array} \quad \begin{array}{r} 454 \\ \times\ 37 \\ \hline \end{array} \quad \begin{array}{r} 527 \\ \times\ 18 \\ \hline \end{array} \quad \begin{array}{r} 505 \\ \times\ 51 \\ \hline \end{array}$$

$$\begin{array}{r} 520 \\ \times\ 27 \\ \hline \end{array} \quad \begin{array}{r} 318 \\ \times\ 45 \\ \hline \end{array} \quad \begin{array}{r} 529 \\ \times\ 54 \\ \hline \end{array} \quad \begin{array}{r} 454 \\ \times\ 25 \\ \hline \end{array}$$

$$\begin{array}{r} 364 \\ \times\ 16 \\ \hline \end{array} \quad \begin{array}{r} 475 \\ \times\ 27 \\ \hline \end{array} \quad \begin{array}{r} 457 \\ \times\ 48 \\ \hline \end{array}$$

$$\begin{array}{r} 356 \\ \times\ 39 \\ \hline \end{array} \quad \begin{array}{r} 465 \\ \times\ 48 \\ \hline \end{array} \quad \begin{array}{r} 590 \\ \times\ 27 \\ \hline \end{array}$$

61) Multiply 3-Digits by 2-Digits

Name:	Data:	Score:

$$\begin{array}{r} 624 \\ \times\ 20 \\ \hline \end{array} \qquad \begin{array}{r} 627 \\ \times\ 43 \\ \hline \end{array} \qquad \begin{array}{r} 634 \\ \times\ 54 \\ \hline \end{array} \qquad \begin{array}{r} 624 \\ \times\ 15 \\ \hline \end{array}$$

$$\begin{array}{r} 504 \\ \times\ 26 \\ \hline \end{array} \qquad \begin{array}{r} 654 \\ \times\ 47 \\ \hline \end{array} \qquad \begin{array}{r} 627 \\ \times\ 38 \\ \hline \end{array} \qquad \begin{array}{r} 605 \\ \times\ 49 \\ \hline \end{array}$$

$$\begin{array}{r} 421 \\ \times\ 64 \\ \hline \end{array} \qquad \begin{array}{r} 512 \\ \times\ 43 \\ \hline \end{array} \qquad \begin{array}{r} 603 \\ \times\ 52 \\ \hline \end{array} \qquad \begin{array}{r} 356 \\ \times\ 35 \\ \hline \end{array}$$

$$\begin{array}{r} 568 \\ \times\ 16 \\ \hline \end{array} \qquad \begin{array}{r} 677 \\ \times\ 27 \\ \hline \end{array} \qquad \begin{array}{r} 506 \\ \times\ 48 \\ \hline \end{array}$$

62) Multiply 3-Digits by 2-Digits

Name:	Data:	Score:

$$\begin{array}{r} 734 \\ \times\ 42 \\ \hline \end{array} \qquad \begin{array}{r} 437 \\ \times\ 53 \\ \hline \end{array} \qquad \begin{array}{r} 714 \\ \times\ 54 \\ \hline \end{array} \qquad \begin{array}{r} 734 \\ \times\ 35 \\ \hline \end{array}$$

$$\begin{array}{r} 764 \\ \times\ 16 \\ \hline \end{array} \qquad \begin{array}{r} 764 \\ \times\ 27 \\ \hline \end{array} \qquad \begin{array}{r} 737 \\ \times\ 48 \\ \hline \end{array} \qquad \begin{array}{r} 715 \\ \times\ 59 \\ \hline \end{array}$$

$$\begin{array}{r} 730 \\ \times\ 37 \\ \hline \end{array} \qquad \begin{array}{r} 628 \\ \times\ 35 \\ \hline \end{array} \qquad \begin{array}{r} 709 \\ \times\ 64 \\ \hline \end{array}$$

$$\begin{array}{r} 651 \\ \times\ 39 \\ \hline \end{array} \qquad \begin{array}{r} 562 \\ \times\ 28 \\ \hline \end{array} \qquad \begin{array}{r} 693 \\ \times\ 47 \\ \hline \end{array}$$

63) Multiply 3-Digits by 2-Digits

Name:	**Data:**	**Score:**

$$\begin{array}{r} 823 \\ \times\ 20 \\ \hline \end{array} \quad \begin{array}{r} 823 \\ \times 43 \\ \hline \end{array} \quad \begin{array}{r} 834 \\ \times\ 34 \\ \hline \end{array} \quad \begin{array}{r} 803 \\ \times\ 15 \\ \hline \end{array}$$

$$\begin{array}{r} 805 \\ \times\ 56 \\ \hline \end{array} \quad \begin{array}{r} 755 \\ \times\ 47 \\ \hline \end{array} \quad \begin{array}{r} 825 \\ \times\ 38 \\ \hline \end{array} \quad \begin{array}{r} 705 \\ \times\ 29 \\ \hline \end{array}$$

$$\begin{array}{r} 824 \\ \times\ 14 \\ \hline \end{array} \quad \begin{array}{r} 814 \\ \times\ 23 \\ \hline \end{array} \quad \begin{array}{r} 704 \\ \times\ 42 \\ \hline \end{array}$$

$$\begin{array}{r} 867 \\ \times\ 56 \\ \hline \end{array} \quad \begin{array}{r} 704 \\ \times\ 75 \\ \hline \end{array} \quad \begin{array}{r} 807 \\ \times\ 48 \\ \hline \end{array}$$

64) Multiply 3-Digits by 2-Digits

Name:	Data:	Score:

$$\begin{array}{r} 936 \\ \times\ 51 \\ \hline \end{array}$$

$$\begin{array}{r} 439 \\ \times\ 23 \\ \hline \end{array}$$

$$\begin{array}{r} 514 \\ \times\ 34 \\ \hline \end{array}$$

$$\begin{array}{r} 937 \\ \times\ 45 \\ \hline \end{array}$$

$$\begin{array}{r} 967 \\ \times\ 26 \\ \hline \end{array}$$

$$\begin{array}{r} 468 \\ \times\ 17 \\ \hline \end{array}$$

$$\begin{array}{r} 535 \\ \times\ 48 \\ \hline \end{array}$$

$$\begin{array}{r} 915 \\ \times\ 59 \\ \hline \end{array}$$

$$\begin{array}{r} 908 \\ \times\ 67 \\ \hline \end{array}$$

$$\begin{array}{r} 407 \\ \times\ 75 \\ \hline \end{array}$$

$$\begin{array}{r} 854 \\ \times\ 45 \\ \hline \end{array}$$

$$\begin{array}{r} 906 \\ \times\ 85 \\ \hline \end{array}$$

$$\begin{array}{r} 904 \\ \times\ 43 \\ \hline \end{array}$$

$$\begin{array}{r} 406 \\ \times\ 54 \\ \hline \end{array}$$

$$\begin{array}{r} 895 \\ \times\ 49 \\ \hline \end{array}$$

65) Multiply 3-Digits by 2-Digits

Name:	Data:	Score:

924 × 32	724 × 43	635 × 45	724 × 35
306 × 26	256 × 47	126 × 38	406 × 49
526 × 54	615 × 23	405 × 62	756 × 43
268 × 93	379 × 91	408 × 98	

66) Multiply 4-Digits by 2-Digits

Name:	Data:	Score:

$$\begin{array}{r} 1034 \\ \times \quad 42 \\ \hline \end{array}$$

$$\begin{array}{r} 2204 \\ \times \quad 53 \\ \hline \end{array}$$

$$\begin{array}{r} 3034 \\ \times \quad 64 \\ \hline \end{array}$$

$$\begin{array}{r} 4254 \\ \times \quad 25 \\ \hline \end{array}$$

$$\begin{array}{r} 5204 \\ \times \quad 56 \\ \hline \end{array}$$

$$\begin{array}{r} 6254 \\ \times \quad 47 \\ \hline \end{array}$$

$$\begin{array}{r} 2213 \\ \times \quad 48 \\ \hline \end{array}$$

$$\begin{array}{r} 1272 \\ \times \quad 29 \\ \hline \end{array}$$

$$\begin{array}{r} 8037 \\ \times \quad 42 \\ \hline \end{array}$$

$$\begin{array}{r} 6276 \\ \times \quad 53 \\ \hline \end{array}$$

$$\begin{array}{r} 7504 \\ \times \quad 64 \\ \hline \end{array}$$

$$\begin{array}{r} 8935 \\ \times \quad 15 \\ \hline \end{array}$$

$$\begin{array}{r} 3204 \\ \times \quad 76 \\ \hline \end{array}$$

$$\begin{array}{r} 3009 \\ \times \quad 57 \\ \hline \end{array}$$

$$\begin{array}{r} 1248 \\ \times \quad 26 \\ \hline \end{array}$$

67) Multiply 4-Digits by 2-Digits

Name:	Data:	Score:

$$\begin{array}{r} 2035 \\ \times \quad 32 \\ \hline \end{array}$$

$$\begin{array}{r} 3205 \\ \times \quad 43 \\ \hline \end{array}$$

$$\begin{array}{r} 4635 \\ \times \quad 54 \\ \hline \end{array}$$

$$\begin{array}{r} 5205 \\ \times \quad 65 \\ \hline \end{array}$$

$$\begin{array}{r} 6205 \\ \times \quad 56 \\ \hline \end{array}$$

$$\begin{array}{r} 7215 \\ \times \quad 67 \\ \hline \end{array}$$

$$\begin{array}{r} 8235 \\ \times \quad 48 \\ \hline \end{array}$$

$$\begin{array}{r} 9134 \\ \times \quad 39 \\ \hline \end{array}$$

$$\begin{array}{r} 1038 \\ \times \quad 92 \\ \hline \end{array}$$

$$\begin{array}{r} 2047 \\ \times \quad 83 \\ \hline \end{array}$$

$$\begin{array}{r} 3035 \\ \times \quad 54 \\ \hline \end{array}$$

$$\begin{array}{r} 4046 \\ \times \quad 91 \\ \hline \end{array}$$

$$\begin{array}{r} 5205 \\ \times \quad 76 \\ \hline \end{array}$$

$$\begin{array}{r} 6538 \\ \times \quad 47 \\ \hline \end{array}$$

$$\begin{array}{r} 7239 \\ \times \quad 58 \\ \hline \end{array}$$

68) Multiply 3-Digits by 3-Digits

Name:	Data:	Score:

$$\begin{array}{r} 124 \\ \times 142 \\ \hline \end{array}$$

$$\begin{array}{r} 127 \\ \times 423 \\ \hline \end{array}$$

$$\begin{array}{r} 104 \\ \times 34 \\ \hline \end{array}$$

$$\begin{array}{r} 224 \\ \times 45 \\ \hline \end{array}$$

$$\begin{array}{r} 204 \\ \times 456 \\ \hline \end{array}$$

$$\begin{array}{r} 154 \\ \times 227 \\ \hline \end{array}$$

$$\begin{array}{r} 127 \\ \times 548 \\ \hline \end{array}$$

$$\begin{array}{r} 205 \\ \times 629 \\ \hline \end{array}$$

$$\begin{array}{r} 240 \\ \times 357 \\ \hline \end{array}$$

$$\begin{array}{r} 118 \\ \times 745 \\ \hline \end{array}$$

$$\begin{array}{r} 129 \\ \times 454 \\ \hline \end{array}$$

69) Multiply 3-Digits by 3-Digits

Name:	Data:	Score:

$$\begin{array}{r} 324 \\ \times 132 \\ \hline \end{array}$$

$$\begin{array}{r} 327 \\ \times 523 \\ \hline \end{array}$$

$$\begin{array}{r} 334 \\ \times 654 \\ \hline \end{array}$$

$$\begin{array}{r} 425 \\ \times 235 \\ \hline \end{array}$$

$$\begin{array}{r} 404 \\ \times 456 \\ \hline \end{array}$$

$$\begin{array}{r} 454 \\ \times 650 \\ \hline \end{array}$$

$$\begin{array}{r} 407 \\ \times 408 \\ \hline \end{array}$$

$$\begin{array}{r} 405 \\ \times 507 \\ \hline \end{array}$$

$$\begin{array}{r} 320 \\ \times 900 \\ \hline \end{array}$$

$$\begin{array}{r} 602 \\ \times 685 \\ \hline \end{array}$$

$$\begin{array}{r} 703 \\ \times 254 \\ \hline \end{array}$$

70) Multiply 3-Digits by 3-Digits

Name:	Data:	Score:

$$\begin{array}{r} 350 \\ \times 461 \\ \hline \end{array} \quad \begin{array}{r} 527 \\ \times 253 \\ \hline \end{array} \quad \begin{array}{r} 504 \\ \times 324 \\ \hline \end{array} \quad \begin{array}{r} 625 \\ \times 235 \\ \hline \end{array}$$

$$\begin{array}{r} 245 \\ \times 142 \\ \hline \end{array} \quad \begin{array}{r} 415 \\ \times 934 \\ \hline \end{array} \quad \begin{array}{r} 527 \\ \times 915 \\ \hline \end{array} \quad \begin{array}{r} 505 \\ \times 604 \\ \hline \end{array}$$

$$\begin{array}{r} 308 \\ \times 608 \\ \hline \end{array} \quad \begin{array}{r} 318 \\ \times 129 \\ \hline \end{array} \quad \begin{array}{r} 850 \\ \times 465 \\ \hline \end{array}$$

71) Multiply 3-Digits by 3-Digits

Name:	Data:	Score:

$$\begin{array}{r} 621 \\ \times\ 420 \\ \hline \end{array} \quad \begin{array}{r} 622 \\ \times\ 543 \\ \hline \end{array} \quad \begin{array}{r} 633 \\ \times\ 654 \\ \hline \end{array} \quad \begin{array}{r} 625 \\ \times\ 715 \\ \hline \end{array}$$

$$\begin{array}{r} 504 \\ \times\ 826 \\ \hline \end{array} \quad \begin{array}{r} 654 \\ \times\ 947 \\ \hline \end{array} \quad \begin{array}{r} 627 \\ \times\ 138 \\ \hline \end{array} \quad \begin{array}{r} 605 \\ \times\ 249 \\ \hline \end{array}$$

$$\begin{array}{r} 421 \\ \times\ 364 \\ \hline \end{array} \quad \begin{array}{r} 512 \\ \times\ 543 \\ \hline \end{array} \quad \begin{array}{r} 603 \\ \times\ 652 \\ \hline \end{array}$$

72) Multiply 4-Digits by 3-Digits

Name:	Data:	Score:

$$\begin{array}{r} 2034 \\ \times\ 241 \\ \hline \end{array}$$

$$\begin{array}{r} 3204 \\ \times\ 352 \\ \hline \end{array}$$

$$\begin{array}{r} 4034 \\ \times\ 463 \\ \hline \end{array}$$

$$\begin{array}{r} 5254 \\ \times\ 524 \\ \hline \end{array}$$

$$\begin{array}{r} 6204 \\ \times\ 655 \\ \hline \end{array}$$

$$\begin{array}{r} 7054 \\ \times\ 746 \\ \hline \end{array}$$

$$\begin{array}{r} 8203 \\ \times\ 647 \\ \hline \end{array}$$

$$\begin{array}{r} 3272 \\ \times\ 428 \\ \hline \end{array}$$

$$\begin{array}{r} 9037 \\ \times\ 942 \\ \hline \end{array}$$

$$\begin{array}{r} 7276 \\ \times\ 853 \\ \hline \end{array}$$

$$\begin{array}{r} 8504 \\ \times\ 764 \\ \hline \end{array}$$

73) Multiply 4-Digits by 3-Digits

Name:	Data:	Score:

$$\begin{array}{r} 4201 \\ \times\ 376 \\ \hline \end{array} \quad \begin{array}{r} 5002 \\ \times\ 257 \\ \hline \end{array} \quad \begin{array}{r} 6243 \\ \times\ 926 \\ \hline \end{array} \quad \begin{array}{r} 7934 \\ \times\ 615 \\ \hline \end{array}$$

$$\begin{array}{r} 8206 \\ \times\ 356 \\ \hline \end{array} \quad \begin{array}{r} 9217 \\ \times\ 267 \\ \hline \end{array} \quad \begin{array}{r} 1236 \\ \times\ 548 \\ \hline \end{array} \quad \begin{array}{r} 2135 \\ \times\ 439 \\ \hline \end{array}$$

$$\begin{array}{r} 3039 \\ \times\ 792 \\ \hline \end{array} \quad \begin{array}{r} 4041 \\ \times\ 683 \\ \hline \end{array} \quad \begin{array}{r} 5035 \\ \times\ 854 \\ \hline \end{array}$$

74) Mixed Review 1

Name:	Data:	Score:

$$\begin{array}{r} 21 \\ \times \ 7 \\ \hline \end{array} \quad \begin{array}{r} 32 \\ \times \ 3 \\ \hline \end{array} \quad \begin{array}{r} 43 \\ \times \ 4 \\ \hline \end{array} \quad \begin{array}{r} 54 \\ \times \ 5 \\ \hline \end{array}$$

$$\begin{array}{r} 54 \\ \times 52 \\ \hline \end{array} \quad \begin{array}{r} 63 \\ \times 64 \\ \hline \end{array} \quad \begin{array}{r} 72 \\ \times 86 \\ \hline \end{array} \quad \begin{array}{r} 81 \\ \times 78 \\ \hline \end{array}$$

$$\begin{array}{r} 74 \\ \times 93 \\ \hline \end{array} \quad \begin{array}{r} 95 \\ \times 26 \\ \hline \end{array} \quad \begin{array}{r} 93 \\ \times 19 \\ \hline \end{array} \quad \begin{array}{r} 89 \\ \times 64 \\ \hline \end{array}$$

$$\begin{array}{r} 48 \\ \times 46 \\ \hline \end{array} \quad \begin{array}{r} 57 \\ \times 29 \\ \hline \end{array} \quad \begin{array}{r} 74 \\ \times 46 \\ \hline \end{array}$$

75) Mixed Review 2

Name:	Data:	Score:

$$\begin{array}{r} 81 \\ \times 12 \\ \hline \end{array} \quad \begin{array}{r} 82 \\ \times 25 \\ \hline \end{array} \quad \begin{array}{r} 73 \\ \times 37 \\ \hline \end{array} \quad \begin{array}{r} 74 \\ \times 8 \\ \hline \end{array}$$

$$\begin{array}{r} 65 \\ \times 50 \\ \hline \end{array} \quad \begin{array}{r} 66 \\ \times 43 \\ \hline \end{array} \quad \begin{array}{r} 89 \\ \times 28 \\ \hline \end{array} \quad \begin{array}{r} 97 \\ \times 57 \\ \hline \end{array}$$

$$\begin{array}{r} 80 \\ \times 17 \\ \hline \end{array} \quad \begin{array}{r} 80 \\ \times 97 \\ \hline \end{array} \quad \begin{array}{r} 86 \\ \times 49 \\ \hline \end{array} \quad \begin{array}{r} 85 \\ \times 68 \\ \hline \end{array}$$

$$\begin{array}{r} 54 \\ \times 56 \\ \hline \end{array} \quad \begin{array}{r} 63 \\ \times 48 \\ \hline \end{array} \quad \begin{array}{r} 62 \\ \times 38 \\ \hline \end{array}$$

76) Mixed Review 3

Name:	Data:	Score:

$$\begin{array}{r} 95 \\ \times\ \ 8 \\ \hline \end{array} \qquad \begin{array}{r} 87 \\ \times\ \ 5 \\ \hline \end{array} \qquad \begin{array}{r} 76 \\ \times\ \ 2 \\ \hline \end{array} \qquad \begin{array}{r} 65 \\ \times\ \ 7 \\ \hline \end{array}$$

$$\begin{array}{r} 64 \\ \times 10 \\ \hline \end{array} \qquad \begin{array}{r} 73 \\ \times 62 \\ \hline \end{array} \qquad \begin{array}{r} 82 \\ \times 34 \\ \hline \end{array} \qquad \begin{array}{r} 95 \\ \times 56 \\ \hline \end{array}$$

$$\begin{array}{r} 94 \\ \times 32 \\ \hline \end{array} \qquad \begin{array}{r} 85 \\ \times 54 \\ \hline \end{array} \qquad \begin{array}{r} 83 \\ \times 15 \\ \hline \end{array} \qquad \begin{array}{r} 79 \\ \times 68 \\ \hline \end{array}$$

$$\begin{array}{r} 78 \\ \times 47 \\ \hline \end{array} \qquad \begin{array}{r} 67 \\ \times 68 \\ \hline \end{array} \qquad \begin{array}{r} 64 \\ \times 51 \\ \hline \end{array}$$

$$\begin{array}{r} 95 \\ \times 24 \\ \hline \end{array} \qquad \begin{array}{r} 64 \\ \times 35 \\ \hline \end{array} \qquad \begin{array}{r} 60 \\ \times 38 \\ \hline \end{array}$$

77) Mixed Review 4

Name:	Data:	Score:

$$\begin{array}{r} 80 \\ \times 20 \\ \hline \end{array} \quad \begin{array}{r} 86 \\ \times 25 \\ \hline \end{array} \quad \begin{array}{r} 76 \\ \times 49 \\ \hline \end{array} \quad \begin{array}{r} 77 \\ \times 38 \\ \hline \end{array}$$

$$\begin{array}{r} 60 \\ \times 47 \\ \hline \end{array} \quad \begin{array}{r} 61 \\ \times 46 \\ \hline \end{array} \quad \begin{array}{r} 82 \\ \times 35 \\ \hline \end{array} \quad \begin{array}{r} 93 \\ \times 24 \\ \hline \end{array}$$

$$\begin{array}{r} 94 \\ \times 43 \\ \hline \end{array} \quad \begin{array}{r} 86 \\ \times 32 \\ \hline \end{array} \quad \begin{array}{r} 85 \\ \times 21 \\ \hline \end{array} \quad \begin{array}{r} 87 \\ \times 42 \\ \hline \end{array}$$

$$\begin{array}{r} 58 \\ \times 14 \\ \hline \end{array} \quad \begin{array}{r} 69 \\ \times 46 \\ \hline \end{array} \quad \begin{array}{r} 61 \\ \times 68 \\ \hline \end{array}$$

$$\begin{array}{r} 92 \\ \times 39 \\ \hline \end{array} \quad \begin{array}{r} 64 \\ \times 52 \\ \hline \end{array} \quad \begin{array}{r} 76 \\ \times 17 \\ \hline \end{array}$$

78) Mixed Review 5

Name:	Data:	Score:

$$\begin{array}{r} 936 \\ \times\ 51 \\ \hline \end{array} \quad \begin{array}{r} 439 \\ \times\ 23 \\ \hline \end{array} \quad \begin{array}{r} 514 \\ \times\ 34 \\ \hline \end{array} \quad \begin{array}{r} 937 \\ \times\ 45 \\ \hline \end{array}$$

$$\begin{array}{r} 967 \\ \times\ 26 \\ \hline \end{array} \quad \begin{array}{r} 468 \\ \times\ 17 \\ \hline \end{array} \quad \begin{array}{r} 535 \\ \times\ 48 \\ \hline \end{array} \quad \begin{array}{r} 915 \\ \times\ 59 \\ \hline \end{array}$$

$$\begin{array}{r} 908 \\ \times\ 67 \\ \hline \end{array} \quad \begin{array}{r} 407 \\ \times\ 75 \\ \hline \end{array} \quad \begin{array}{r} 854 \\ \times\ 45 \\ \hline \end{array} \quad \begin{array}{r} 906 \\ \times\ 85 \\ \hline \end{array}$$

$$\begin{array}{r} 904 \\ \times\ 43 \\ \hline \end{array} \quad \begin{array}{r} 406 \\ \times\ 54 \\ \hline \end{array} \quad \begin{array}{r} 895 \\ \times\ 49 \\ \hline \end{array}$$

79) Mixed Review 6

Name:	Data:	Score:

$$\begin{array}{r} 821 \\ \times\ \ 2 \\ \hline \end{array} \qquad \begin{array}{r} 822 \\ \times\ \ 3 \\ \hline \end{array} \qquad \begin{array}{r} 733 \\ \times\ \ 5 \\ \hline \end{array} \qquad \begin{array}{r} 625 \\ \times\ \ 5 \\ \hline \end{array}$$

$$\begin{array}{r} 406 \\ \times\ 25 \\ \hline \end{array} \qquad \begin{array}{r} 357 \\ \times\ 40 \\ \hline \end{array} \qquad \begin{array}{r} 228 \\ \times\ 31 \\ \hline \end{array} \qquad \begin{array}{r} 509 \\ \times\ 47 \\ \hline \end{array}$$

$$\begin{array}{r} 621 \\ \times\ 53 \\ \hline \end{array} \qquad \begin{array}{r} 712 \\ \times\ 27 \\ \hline \end{array} \qquad \begin{array}{r} 503 \\ \times\ 65 \\ \hline \end{array} \qquad \begin{array}{r} 854 \\ \times\ 45 \\ \hline \end{array}$$

$$\begin{array}{r} 365 \\ \times\ 94 \\ \hline \end{array} \qquad \begin{array}{r} 476 \\ \times\ 92 \\ \hline \end{array} \qquad \begin{array}{r} 507 \\ \times\ 91 \\ \hline \end{array}$$

Wow! To the Finish line!

Color it using Your Favorite Colors

Get your certificate of success for Multiplications from your teacher!

www.ingramcontent.com/pod-product-compliance
Lightning Source LLC
Chambersburg PA
CBHW081209180526
45170CB00006B/2268
* 9 7 8 1 5 4 2 4 3 4 1 9 5 *